Making Progress in Housing
A framework for collaborative research

This book presents a new approach to housing research, one that is relevant to all the social sciences.

Housing research is diverse and operates across many disciplines, approaches and methods making collaboration difficult. This book outlines a methodological framework that enables researchers from many different fields to collaborate in solving complex and seemingly intractable housing problems. It shows how we can make progress in housing research and deliver better housing outcomes through an integrated approach.

Drawing on the work of renowned Canadian methodologist, philosopher, theologian and economist, Bernard Lonergan (1904–1984), McNelis outlines a framework for collaborative research: Functional Collaboration. This new form of collaboration divides up the work of housing research into functional specialties. These distinguish eight inter-related questions that arise in the process of moving from the current housing situation through to providing practical advice to decision-makers. To answer each question a different method is required. *Making Progress in Housing* is the result of finding new answers to this complete set of eight inter-related questions.

This approach to collaboration opens up a new discourse on method in housing and social research as well as new debates on progress and the nature of science.

Sean McNelis has over 30 years' experience in housing management, housing policy and housing research. He is a research fellow at The Swinburne Institute for Social Research, Swinburne University of Technology, Australia.

Making Progress in Housing

A Framework for
Collaborative Research

Sean McNelis

Dialectic   Foundations

History   Policies

Interpretation   Systematics

Research   Communications

Routledge
Taylor & Francis Group

LONDON AND NEW YORK

First published 2014 by Routledge

2 Park Square, Milton Park, Abingdon, Oxfordshire OX14 4RN
711 Third Avenue, New York, NY 10017

Routledge is an imprint of the Taylor & Francis Group, an informa business.

First issued in paperback 201

British Library Cataloguing in Publication Data
A catalogue record for this book is available from the British Library

Library of Congress Cataloging in Publication Data
McNelis, Sean.
Making progress in housing : a framework for collaborative research / Sean McNelis.
pages cm
Includes bibliographical references and index.
ISBN 978-0-415-70346-8 (hbk. : alk. paper) -- ISBN 978-1-315-83260-9 (ebk)
1. Housing--Research. I. Title.
HD7287.M426 2014
363.5072--dc23
2013029021

ISBN: 978-0-415-70346-8 (hbk)
ISBN: 978-0-367-00122-3 (pbk)

Typeset in Times
by Saxon Graphics Ltd, Derby

To Brenda, Eamon, Jesse and Maeve

You can have teamwork insofar, first of all, as the fact of reciprocal dependence is understood and appreciated. Not only is that understanding required; one has to be familiar with what is called the *acquis*, what has been settled, what no one can doubt at the present time. You're doing a big thing when you can upset that, but you have to know where things stand at the present time, what has already been achieved, to be able to see what is new in its novelty as a consequence. There is a necessity of easy and rapid communication. The point to the university, the research school, the publications, the periodicals, the books, the congresses, is the easy and rapid communication of people working in specialized fields which are reciprocally dependent.

Bernard Lonergan ([1968] 2010: 462)

Contents

Abbreviations

ABS	Australian Bureau of Statistics
ACOSS	Australian Council of Social Service
ACT	Australian Capital Territory
AHURI	Australian Housing and Urban Research Institute
AIHW	Australian Institute of Health and Welfare
CSHA	Commonwealth-State Housing Agreement (Australia)
HCV	Housing Commission Victoria
NSW	New South Wales (Australian State)
SAHT	South Australian Housing Trust
SCRGSP	Steering Committee for the Review of Government Service Provision
SHA	State Housing Authority
SHO	Social Housing Organisation
SHP	Structures of Housing Provision
VCOSS	Victorian Council of Social Service

Acknowledgements

Thank you to my colleagues at the Swinburne Institute for Social Research who create a congenial atmosphere for research, in particular to Professor Terry Burke for his comments on early drafts.

Over the past ten years, I have appreciated the support and encouragement of members of the Melbourne Lonergan Group who, even now, are excited and enthralled by the richness of Lonergan's thinking. They have given me an opportunity to present my thinking, at times when I was struggling to articulate my thoughts and at other times when my insights still needed further refinement: John Little, Stephen Ames, Jamie Pearce, John Boyd-Turner, Tom Daley, Robin Koning, Chris White and Tony McSweeney. My thanks also to Peter Beer from the Sydney Lonergan Centre for his hospitality during many hours of reading and research at the Centre. At difficult times, when it just got too hard, Tom Halloran was always happy to discuss my latest thoughts.

My thanks to Peter Fowler for drawing the figures and for re-drawing versions of the 'Functional Collaboration' illustration from the website of the Society for the Glocalization of Effective Methods of Evolving (SGEME) (www.sgeme.org) and to Sandy Gillis, editor of Axial Publishing, for permission to use this illustration.

Feedback and comments

I would welcome feedback and comments. These can be left on the forum at: www.artfulhousing.com.au

Preface

For over thirty years, I have been involved in housing management, housing policy and housing research: as a housing manager living on a high-rise public housing estate in Melbourne, Australia; as a community development worker developing and managing different types of housing organisations; as a founding member, director and tenant of a small housing co-operative; as a housing policy worker with VCOSS (a peak community organisation), analysing government housing policy and advocating for better housing policies; as a researcher with Ecumenical Housing, a small but influential community housing organisation; and, more recently, as a housing researcher at the Swinburne Institute for Social Research (within Swinburne University of Technology).

Seared on my mind through these years of housing work is the anger, bewilderment, trauma and frustrated hopes of many people seeking but unable to find adequate housing for themselves and their families. On the other hand, there is the joy, the freedom, the new hopes, the opening up of new possibilities as others found housing that not only provided shelter but also a place within which they and their families could create a home for privacy and intimacy, create a space for self-expression and growth, and create a base from which to participate in their local and broader communities.

The consequences of housing practices are writ large in the stories of the most vulnerable members of our community. These stories bear witness to the strength of the human spirit, to the value and fragility of human life and to the importance of collaboration for survival, for meaning and for development. They also reveal the inadequacy of our social institutions and their failure to respect, protect and support all of its members. Those on the margins are deprived of sufficient resources to participate in and influence not just their local environment but also to shape our society. Insofar as current housing practices contribute to their exploitation, disenfranchisement and alienation, they are an affront to common human decency, to democracy, to the right of each person to a place they feel is their home and to a place where they belong. They are an affront to our common human solidarity and our social responsibility as custodians of the land and of the world's resources.

It is stories of these vulnerable members of our community that continue to motivate my housing research (as they do many housing researchers). At the same time, however, it has been the inadequacy of housing research that has contributed to our current inability to analyse housing situations, to ask the more difficult questions and to think through issues; our failure to provide practical and innovative advice to decision-makers; our bowing to the interests of groups seeking short-term fixes rather than sound long-term solutions. No doubt many housing researchers have done excellent research and proposed good policies, but our common-sense framework for analysis and our often partisan opinions allow decision-makers to dismiss this work too easily as 'just one point of view' while their own vested interests remained concealed.

In many ways this book is a reflection upon, a critique and an evaluation of my own housing research[1] just as much as it is of housing research more generally. It arises out of my dissatisfaction with this research.

The current culture of housing research is founded on everyday common sense, a search for immediate short-term fixes to difficult problems. One sign of this approach is the ever-present question of the usefulness of any research and its practical implications. Over many years I have become increasingly dissatisfied with this style of research. In my view, this dominant and largely unreflective culture is no longer adequate to the task. Already some researchers are aware of this. Over twenty years ago, Jim Kemeny called for theory in housing research, for relating housing to debates in the social sciences and for a greater sense of reflexivity, in particular, by examining the epistemological grounds of what we are doing (among them the questions we pose) (Kemeny 1992: xvii). I would push these further. While the shift to theory is critical, so too is the need for a better understanding of what theory is (King 2009: 51). While we need to relate housing to debates in the social sciences, we also desperately need some way of integrating and relating the disparate array of disciplines and their various methods. While we need to acknowledge, put forward and justify the epistemological grounds of what we are doing, we need to do more than just that. We need to find some basis upon which we not only resolve conflicts regarding epistemological grounds but also explain why others put forward contrary epistemological grounds: 'If Descartes has imposed upon subsequent philosophers a requirement of rigorous method, Hegel has obliged them not only to account for their own views but also to explain the existence of contrary convictions and opinions' (Lonergan [1957] 1992: 553).

This book is proposing a scientific approach to housing research. It proposes a framework which will facilitate collaboration among researchers operating in a diversity of disciplines using a diversity of methods; a framework through which we can make progress in housing and which offers some hope for the future. It is not an approach of my own making. It draws its inspiration from the work of Bernard Lonergan (1904–1984), a Canadian methodologist, philosopher, theologian and economist.

The context of Lonergan's work was primarily theology and it may seem strange that the writings of a theologian should have any relevance to housing research. Some may view this as taking a little too far Jim Kemeny's call (1992: 11–12) for housing researchers to 'return to their parent disciplines and reconceptualize housing according to the theories and concepts prevalent in each area' (King 2009: 44). Strange as it may seem, I have come to the conclusion that Lonergan's insights are not just relevant but extremely important to the future of housing research.

In 2002 when I first began research work in an academic context, I was soon confronted by a vast array of different types of housing research and by the many – and at times acrimonious – debates between different researchers. It seemed to me, however, that these debates were often at cross-purposes because the researchers were doing different things. In reaching this conclusion, I was recalling work by Lonergan on functional specialisation. I had first encountered his writings in the early 1970s and then later on in the early 1980s as part of a Bachelor of Theology. I felt, however, that it demanded a more intensive investigation and I resolved to work on it amid my other responsibilities.

While reading Lonergan's key works, *Insight: A Study of Human Understanding* ([1957] 1992) and *Method in Theology* ([1972] 1990), I had found something deeply challenging and attractive. For me Lonergan posed, and continues to pose, a personal challenge, a challenge to authenticity and a challenge to deal with things at a deeper level. As a Catholic philosopher and theologian of the twentieth century, he had cast off the decadent scholasticism he had inherited and began a dialogue between the authentic tradition of Aquinas and contemporary science, mathematics, logic and philosophy. He was not content with simplistic answers. Like the best theologians, he explored the contemporary questions of his day. He understood his faith as consistent with human understanding and reason. He was not simply concerned with method in theology but rather with the unfolding of world process itself in its physical, chemical, biological, psychological, intelligent and religious dimensions. Lonergan was a searcher, not just for an understanding of our world, but for ways in which to make living better for all.

My attraction to Lonergan stems from his critical appreciation of all serious thinkers and from the way in which he dealt with differing viewpoints. He was not content with recognising and tolerating conflicts and differences, nor with forceful and fearful rejection and opposition to alternative views. Such conflicts and differences raised key unresolved questions for him personally. His *modus operandi* was one of integration. He took varying and conflicting viewpoints and reached for a new higher viewpoint, one which integrated the contribution of other viewpoints, and one which sorted out and appreciated their contribution while recognising their limitations.

This book is my interpretation of Lonergan's book *Method in Theology*. However, it is an interpretation in the context of housing research, not of theology. Consequently it pivots between my insights into his seminal discovery of functional specialties as elaborated in *Method in Theology* and my insights into the orientation of housing researchers who are embedded within the everyday world of common-sense living and a particular taken-for-granted research context. My primary purpose is to communicate to housing researchers a new understanding of science as Functional Collaboration (a phrase not used by Lonergan but which captures his intent). My purpose is not a faithful (and detailed) exegesis of his writings. Rather, I seek to develop his central theme of the dynamic operations of the subject as subject as reflected in the question: what am I doing when I am doing housing research? I would not claim to have adequately interpreted his writings on the functional specialties nor his broader philosophical, theological and economic writings. *Caveat emptor!* (Let the reader beware!) In attempting to communicate Functional Collaboration to housing researchers as a new understanding of science, I have had to make decisions about what to include and what to exclude. So, I have deliberately steered away from using Lonergan's terminology except in some key areas and have not elaborated on some areas which many would regard as central to his work. In my view, to do otherwise would unnecessarily complicate the discussion and distract the reader from the core issue of what we are doing when we are doing housing research. However, I hope that the Lonergan scholar may forgive such lapses while recognising how I have transposed the key insights of Lonergan into the context of housing research.

As I began more intensive work on functional specialisation – some ten years ago now – Lonergan pointed me to something beyond my current understanding of science. He proffered an invitation. My beginning was simply a matter of belief that concerted work on Lonergan's writings would take me into a new world, open up new vistas and new horizons. He, like many scientists, writers, visionaries and social activists, is a prophet pointing to something better. He, like many others, invites us: to become better persons; to take up new challenges; to become cognisant of the injustices we perpetrate and feel the pain of disappointments and sorrows in those around us; to stop and enjoy the moments of laughter and of achievement of ourselves and those around us; to ask new questions and seek new answers; and to love more deeply and intimately. Only after ten years' solid work as I 'retrace my steps' can I appreciate that I have reached a point where I'm beginning to understand something of what he is offering.

In this book I am raising a fundamental question about the current culture of housing research. I am critical of this culture and point to the need for its radical transformation if it is to provide practical and innovative advice to decision-makers and make progress in housing.

As a housing researcher enmeshed in a common-sense framework, I found reaching some minimal understanding of Functional Collaboration (and each functional specialty) a major challenge; I found it very difficult to imagine, to fantasise about something which requires a fundamental transformation of my thinking and my doing of housing research. I found it made demands upon both self-understanding and upon decisions I make as to who I will become both as a housing researcher and as person. William Mathews (2005: 151) in his intellectual biography of Lonergan notes: 'It is important that, at this beginning, we as interpreters try to recognize in ourselves what Lonergan is talking about. Otherwise we will not be able to begin, will fail to gain an entry into his journey and path.' My challenge throughout the long gestation of this book stemmed from the great difficulty I faced in grasping who I am and what I was doing.

I had to come to some understanding of my practices as a housing researcher, then some understanding of what I was doing when I was evaluating them, and finally some understanding of what I was doing when I decided to implement something new.

There are no certainties as to the success of Functional Collaboration as a scientific approach to housing. Housing research continually faces the problem of the vested interests of researchers, decision-makers, funders, investors, land and building industries, politicians, bureaucrats and others. The scientific approach proposed here includes a process for continually offsetting these vested interests. By confronting them and by revealing their inconsistencies and limitations, the probabilities of success will be improved. Moreover, this scientific approach will draw upon the resources (and operative solutions) in many different societies, cultures and research traditions, it will sort out their contribution to history, it will provide practical advice to decision-makers and, slowly but cumulatively, it will address the seemingly intractable situations that confront housing today.

After thirty years of involvement in housing management, housing policy and housing research, I still hope for and dream of a better future for households seeking adequate housing. I am exploring a radical new understanding (and doing) of housing research. In *The Poetics of Space* (1969: 61), the French philosopher Gaston Bachelard wrote:

> Sometimes the house of the future is better built, lighter and larger than all the houses of the past, so that the image of the dream house is opposed to that of the childhood home. Late in life, with indomitable courage, we continue to say that we are going to do what we have not yet done: we are going to build a house.

I hope I can contribute to the building of a new home for housing research. Just as building a home requires co-operation between many different people, housing research requires a global 'framework for collaborative creativity' that will continually offset partisan interests, provide practical and innovative advice to decision-makers and bear fruit in a better housing system. I hope that housing researchers will have the courage to leave the 'childhood home' of common sense and build the dream home of the future – better built, lighter and larger.

Sean McNelis
July 2013

Notes

1 This includes: various submissions on Australian and Victorian government housing policies (VCOSS 1989, 1990, 1991a, 1991b, 1992, 1993a, 1993b); papers envisaging a future for social housing (McNelis 1992, 1993), reviewing performance monitoring (McNelis 1996) and discussing the future of high-rise public housing (McNelis & Reynolds 2001); AHURI reports on a private investment vehicle for community housing

(McNelis et al. 2001, 2002a, 2002b), on independent living units for older people with low incomes and low assets (McNelis & Herbert 2003; McNelis 2004), on rental systems (McNelis & Burke 2004; McNelis 2006) and on older persons in public housing (McNelis 2007; McNelis et al. 2008); local government housing strategies (Kliger & McNelis 2003a, 2003b; McNelis, Esposto & Neske 2005a, 2005b; McNelis et al. 2005a, 2005b, 2005c; McNelis, Burke & Neske 2006); papers on housing and disability (McNelis & Nichols 1997; Reynolds et al. 1999; McNelis et al. 2002) and on asset management (Ecumenical Housing et al. 1999).

Introduction

Industrialisation in the late eighteenth and early nineteenth centuries in Western Europe saw the migration of thousands of people into cities, the creation of slums with their appalling housing conditions and the emergence of 'the social question'. This was the era when liberal capitalism dominated. One initial response in the mid to late nineteenth century was to mandate minimal housing standards to improve the health of those living in slums and to prevent the spread of disease (Daunton 1991). Other responses emerged from housing reformers such as Octavia Hill (1883), city planners such as Ebenezer Howard's Garden City movement (1902) and philanthropists and employers concerned about the plight of the poor. By the beginning of the twentieth century, governments began to play some role: local councils directly constructing and providing housing; central governments providing funds to support the initiatives of local councils, philanthropists, co-operatives and other types of organisations. In this way some form of social housing emerged in European countries as a 'politically acceptable solution' to the housing problem. This solution received a further boost as a result of the Great War, the Great Depression and the Second World War (Harloe 1995).

Nevertheless, despite the emergence of social housing in more developed countries, we continue to confront apparently insurmountable housing problems such as our ongoing inability to deliver housing that is affordable, of good quality, appropriate and sustainable. Even more disturbing are the appalling housing conditions of those living in the slums of Asia, Africa and South America and the ghettos of Europe and North America, and of displaced peoples in refugee camps.

The advent of social housing marks some progress in providing adequate housing for all people. Whether and how long current forms of social housing or, indeed, the private rental market or owner-occupied housing will continue is impossible to predict. Social housing is an achievement which plays an important role in more developed countries. Yet it is a solution with many shortcomings. Current solutions to the problems of inadequate, inappropriate, unaffordable and unsustainable housing are only interim achievements that provide the grounds for some new and better future solution. Housing must continually adapt to the demands of a changing environment. Yet, despite the pressing needs of many people, progress towards new solutions is haphazard and random.

In large part, the development of new solutions depends upon housing research. It is through understanding and evaluating/critiquing the experience of current forms of housing as well as through our creative intelligence and commitment that we discover more adequate solutions. Housing research is not simply a matter of investigating some housing issue and proposing solutions, nor is it simply about analysing policies and recommending new ones, nor is it simply about critiquing individual and collective decisions, nor is it simply about acting strategically and practically. It demands all of these and more. If we are to have something better than haphazard and random developments in housing, we need to put in place a mode of housing research that links these different aspects of research together in such a way that they promote a continuous flow of improvements, a flow that will be ongoing and cumulative.

Housing research is diverse, operating across many disciplines (such as economics, politics, sociology, design, architecture, planning, cultural studies and philosophy), many approaches (such as positivist, social constructionist and critical realist) and many methods (such as qualitative, quantitative, hermeneutical, historical and critical). Diversity is highly valued among researchers. Different perspectives can raise challenging and fundamental questions. On the other hand, diversity fragments housing research with researchers having little sense of how their research relates to the work of others. Fragmentation makes interdisciplinary work difficult as researchers specialise and operate with different presuppositions. Fragmentation promotes silos between different approaches and methods; researchers misunderstand one another and operate at cross-purposes.

Collaboration allows researchers to divide up the work on complex housing problems. For effective collaboration, however, housing researchers need a framework which will hold the diversity of disciplines, approaches and methods together; they need a framework within which researchers from many different contexts and specialties can draw on their strengths and make some contribution to resolving problems.

In this book I will argue that Functional Collaboration is this holistic framework. It is a 'framework for collaborative creativity' (Lonergan [1972] 1990: xi) that divides up the work of housing research more effectively. Rather than focusing on the products of housing research, however, it focuses on questions, the answers they anticipate and the methods by which they are answered. It divides up the work of housing research into functional specialties which distinguish the questions that arise in the process of moving from the current housing system to the creation of something new. Functional Collaboration seeks to: effectively address the situation of everyday living, the complex manifold of events associated with housing; adequately understand these events; identify the different trajectories of housing within different societies and cultures; identify and critically evaluate conflicting movements and groups, conflicting understandings and conflicting policies within housing, and thus reveal the limitations and address the practices of dominant groups; transform the basis upon which the future is constructed; envisage directions or policies which are new to housing ; strategically locate these new housing directions and policies within the larger context of existing technological, economic, political and cultural structures, institutions and capacities; and propose particular changes in the housing practices of individuals and institutions.

Functional Collaboration puts forward a new understanding (and doing) of science, one which is oriented towards making progress in housing through a framework for collaborative research. In this scientific approach, housing researchers specialise in different methods and collaborate by making a specific contribution to the understanding of housing and the implementation of new solutions by presenting decision-makers (whether individuals, businesses, social housing organisations or governments) with practical advice. Functional Collaboration is transdisciplinary, transcultural and global in scope, drawing on the resources of all countries, societies and cultures to address local problems and issues.

Part I introduces the problems, questions and issues faced by housing researchers and decision-makers as they seek to make progress in housing. It introduces two problems: the fragmentation of research and its failure to provide practical advice to decision-makers; and the limitations of the personal and communal horizons of researchers and decision-makers such that they promote interests that aggrandise themselves or their group, i.e. their class, gender, race, nationality etc. It also introduces Functional Collaboration locating its remote origins in the writings of Bernard Lonergan. It briefly notes some examples of its emergence in different fields as authors seek to structure their disciplines by dividing up their work either spontaneously or deliberately into functional specialties. It concludes by reflecting on the different types of questions we ask and by suggesting that a scientific approach to housing will distinguish a complete set of eight inter-related questions: an empirical question, a theoretical question, a historical question, an evaluative/critical question, a transformative/visionary question, a policy question, a strategic question and a practical question. Each question and the method by which it is answered form the basis for a functional specialty.

Part II works through each of the eight functional specialties that constitute Functional Collaboration. It elaborates on a more precise understanding of each question and the method by which each is answered. It envisages a radical change in our understanding of what it is that we are doing when we are doing housing research. It entails a complex series of paradigm shifts and a radical restructuring of the way in which housing research operates.

Part III discusses Functional Collaboration as a whole. It shows how these questions and functional specialties are related to one another. It shows that Functional Collaboration is a scientific approach that relates and integrates disparate methods and types of housing research presenting decision-makers with practical advice on future directions. It gives a more precise understanding to making progress in housing through a framework for collaborative research.

Part I
Context

1 Problems, questions and issues

> Insofar as there is to be a resolute and effective intervention in the historical process one has to postulate that the existential gap must be closed. In other words, one has to postulate that the people who are seeking to influence history, to put their lever at the vital point in historical process, are not operating, not doing their thinking, planning, and policy-making, from within the pair of blinkers of a personal or communal horizon. They have to be people in whom the horizon is coincident with the field. If they are not, then all they possibly can do is increase the confusion and accelerate the doom.
>
> Bernard Lonergan in *Horizon, History, Philosophy* ([1957] 2001c: 306)

Social housing, as the most recent significant achievement in the history of housing, illustrates many of the problems, questions and issues that confront us as we seek to make progress in housing.

Social housing: an overview

Social housing is constituted by a range of elements such as the acquisition of land and dwellings, the design and construction of dwellings, capital financing and operating finance, eligibility and allocation of households, maintenance, asset management and tenancy management (Ball et al. 1988: 5; Burke 1993: 25–33; Paris et al. 1993). These are drawn together in such a way that social housing meets particular purposes such as affordability, equity, efficiency and operational autonomy (Bramley 1991; see also Burke 1993: 34–37; McNelis 2000; Yates c.1994). Each element is structured in such a way that it contributes to and complements the others. Each element is brought about by an aggregate of regular, repetitive activities of many different people. As such, social housing is a collaborative enterprise (Melchin 1991, 1994, 2003) – it is achieved by structuring a broad and diverse range of inter-linked, repetitive and taken-for-granted activities.

Social housing is one enterprise among a very large range of enterprises which meet the material needs of households. It is provided within a larger context of a technology (which provides the know-how for its acquisition, construction, financing and management), an economy (through which many enterprises provide a standard of living), a political system (through which agreement is reached on what is done and how it is done) and a culture (an inherited set of meanings and values that inform our way of living). Social housing achieves its specific purposes by drawing upon, adapting and particularising the operative, taken-for-granted ways of doing things (habits, routines and structured activities) and producing things in this larger context.

Social housing is not static, however. It continues to evolve. It is ever changing and there is a recurrent need to adapt to changing circumstances, to take up opportunities as they arise and to deal with threats as they emerge. The need for change in some area of social housing, however, can be distinguished from the need to reach some agreement about what change will be implemented. Not only is there a recurring need to change the elements of social housing and their particular form, there is also a recurring need to reach agreement about what changes will occur.

Between the inception of social housing and now, there is not only a history of change but also a history of decision-making. In part this history can be traced through the decisions made by governments and/or Social Housing Organisations (SHOs) (depending upon the extent of their autonomy). Each decision had its context, a series of preceding decisions and actions by a broad range of players – governments, political parties, SHOs, managers, practitioners, tenants, applicants, advocates, community activists, researchers, architects, planners, builders, tradespersons, real estate agents, land developers, housing educators etc.

Each player had a different view on the solution to the housing problem. For the most part, all were predominantly interested in immediate short-term solutions. They made their decisions in different ways, however: some were oriented towards improving social housing; some deliberately sought, lobbied and advocated for decisions that furthered their interests and resisted decisions that were not.

Each defined a problem in their own way, exerting their influence and power to ensure that others defined the problem in this way. Differing views (and the decisions that stemmed from them) were the source of conflict between players. For the most part, many individuals and groups sought to exercise their power in their own personal interest or in the interest of some group or other. The role of social housing within society, as indeed all aspects of society, was a site for different and conflicting views: some regarded it as interference in a free market and sought to restrict its application, or even its abolition; some regarded it as the only solution to the housing problem and sought to expand its range, even proposing the abolition of the private market; others viewed social housing as complementary to both owner-occupied housing and the private rental market. As a result, conflicts emerged about the particular elements that constitute social housing (eligibility and allocations, rents, housing stock, finance, legislation etc.) and about what was to done and how it was to be done. As the interests of individuals and groups waxed and waned, and as dominant groups sought to maintain their interests, so too did the particular form of the elements that constituted social housing change (Kemeny 1981; Ball 1983; Jacobs et al. 2004; Dodson 2006; Lawson 2006).

Decision-making is central to the process of change in social housing. While there may be different and even conflicting views as to what is the best course of action, decision-makers seek to bring about something better, to bring about some improvement, to bring about progress in social housing or to use it to make progress in housing more broadly or in some other area of social life. Governments have their view as to what is better; SHOs have their view; tenants and applicants have their view; tenant, community and advocate organisations have their view; academics have their view; local neighbourhoods have their view. And even within these different groups, individuals or sub-groups have their own view as to what is better. We are left, therefore, with the question: have the changes as a result of decisions in the past brought about progress in social housing or have they made it worse? Or, more broadly, have they brought about progress in housing or made it worse?

By making progress a core issue for housing, we immediately face an array of objections. Who is to say whether some change has brought about progress or not? On whose criteria can we assess it? Are general criteria relevant to local concrete circumstances? Are we talking about short-term progress or long-term progress? Are not all decision-makers formed by their own culture and language, their own society, their own socio-economic interests, and thus are in some way biased towards decisions made in their own interests or that of the cultural and socio-economic groups to which they belong? Indeed, is it not simply a matter of who had the power to effect changes in their own interests over and above the interests of other persons or groups?

Those decisions were made in the past. What of the future? Decision-makers now face a different set of issues. Progress, whether some major or minor development in social housing, depends upon our capacity to make changes. What changes we make will in turn depend upon some strategy for effecting those changes, some articulation of what we want to achieve through them, some envisioning of new directions, new policies within a given particular context. It calls for dreams or visions about what the future could be, as well as practicality about what changes have to occur to bring about this near or long-term future. Such change is not simply about the vision and practicality of one person but of a group of people. Where, then, lies the origin and source of a vision for the future? How does this become the vision of a group? What array of persons need to change what they are currently doing, change their habits, change their taken-for-granted way of doing things in order to realise this vision? How are they to integrate this new way of doing within the vast array of things they already do?

Moreover, progress is not simply a matter of continually moving forward. Rather, in seeking to move forward we are confronted by stupidities, greed, entrenched interests and corruption; we are confronted by a current situation in which much housing (including social housing) is poorly designed, poorly located, poorly maintained and environmentally unsustainable. For many households, particularly tenants, their housing does not meet their needs, is not appropriate, overcrowded and unaffordable etc. Indeed, too many people are homeless. Yet, the implementation of practical ideas are distorted and blocked, not only by deliberate self-interest and egoism but also by unacknowledged assumptions which aggrandise particular groups. Many unknowingly support and maintain an unjust status quo: 'Individuals may have the best will in the world, may be good and upright, and yet by their actions contribute to social and historical processes which oppress and dehumanize' (Lamb 1982: 3).

We hope for something better but that hope is often strained and even abandoned in the face of continuing failure and disillusionment. Not only is there a need for progress but also of some remediation, some reversal of the destruction. Sustained progress in housing is not easy.

Housing research and making progress in housing

Decision-makers may regard themselves as competent to make a decision about social housing. They may have some vision on its future or some clear view on what changes they want to implement. They may have accumulated sufficient understanding of its role and how it works. On the other hand, they may have many questions before they reach a decision. Have they properly understood some problem that confronts them? For whom is it a problem? Is their perception of the problem simply a reflection of their own personal interests or the interests of the cultural and socio-economic groups to which they belong? How do they deal with different perspectives on the problem, particularly when they depend upon others (who view it differently) to make changes that will solve the problem? Is the problem they face the real problem or just a manifestation of a deeper problem? Do they have a sufficient grasp of the details to propose a change that will adequately deal with the problem? Is the solution the right one? Indeed, is there only one solution? Will the proposed solution create more problems in the future? These are questions that all responsible decision-makers face.

They may want answers to some or many of the questions they have, for 'in whatever area of experience we are involved, without understanding we are blocked, we cannot go on, nor project our way forward' (Mathews 1987: 245). They may want answers to questions about the current and possible future role of social housing, about what is happening within social housing or impacting on it, and about what to do. Alternatively, they may want answers which will, more or less, justify their decision about what to do; they may want answers that accord, more or less, with their view as to what is happening.

Thus, decision-makers may seek views, information and advice from a variety of sources – colleagues, managers, practitioners, tenants, advocates, consultants, experts, academics etc. – to supplement their understanding. In this way, housing researchers may play a key role in the decision-making process.

Many housing researchers share the view that housing research has a role in policy development but their relationship is often a difficult and complex one (Jones & Seelig 2004, 2005). Housing researchers want to maintain their independence and autonomy. Yet, at the same time, they often want to influence decisions. Decision-makers, while relying upon some types of housing research, often bemoan the relevance of other types of research to their decision-making and even ignore this research because it is not immediately relevant or does not fit with their particular view.

Nevertheless, the role of housing researchers is not limited by the particular expectations of decision-makers. They can serve the process of decision-making by asking and answering questions that decision-makers don't ask or don't want to ask. They may even ask questions about the basis upon which decisions are made, about the values or frameworks or interests of decision-makers. Housing researchers, as researchers, push the questioning process further. They serve decision-making not only by investigating and understanding what is happening but also by working out what could be done to solve problems, by challenging the understanding of decision-makers and by challenging the options they are considering.

The corpus of housing research consists of a very broad range of works: focusing on different aspects of housing – design, technology, economics, social relations, politics, culture, tenure, finance, aesthetics, geography, housing policy, housing discourses etc.; originating in different countries and different cultures; originating from different sources – academia, housing organisations, tenant and community advocates, government and government authorities, housing managers and practitioners. For instance, it ranges across:

- Works from different academic disciplines:
    - sociology (Kemeny 1992; Clapham 2002, 2004, 2005)
    - politics (Bengtsson 1995, 2009a)
    - economics (Maclennan & More 1997; Gibb 2002, 2009)
    - critical geography (Blunt & Dowling 2006)
    - cultural studies (King 2004, 2005, 2008)
    - architecture (Turner 1976; Oliver 2003)
    - philosophy (Bachelard 1969; Heidegger [1951] 1975).
- Works within these disciplines from different epistemological and onto-logical approaches:
    - social constructionist (Kemeny 1992; Clapham 2004; Jacobs et al. 2004)
    - critical realist (Lawson 2001, 2002, 2006, 2010)
    - radical empiricist (Dodson 2007)
    - positivist (most current housing researchers)
    - phenomenologist (Bachelard 1969; King 2004, 2005, 2008).
- Works using different methods:
    - quantitative (UK Department of Communities and Local Government 2012; Dol & Haffner 2010; AIHW 2009; SCRGSP 2012)
    - critical discourse analysis (Marston 2000, 2002, 2004)
    - historical (Harloe 1995; Howe 1988a; Hayward 1996)
    - evaluative (Milligan et al. 2007)
    - critical (Kemeny 1981, 1983; Ronald 2008; Jacobs et al. 2010a)
    - comparative (Ambrose 1992; Dalton 2009)

- transdisciplinary (Lawrence 2004)
- foresight (Burke et al. 2004; Burke & Zakharov 2005).
- Works discussing very different aspects of or concerns about housing:
  - tenure (Lawson 2006; Hulse 2003, 2008)
  - access (Hulse et al. 2006; Hulse et al. 2007)
  - allocations (Burke & Hulse 2003; Hulse & Burke 2005)
  - finance (Hills 1991; Gibb 2002; Hall & Berry 2004, 2007a, 2007b, 2009)
  - rent (Bramley 1991; Grey, Hepworth & Odling-Smee 1981; Yates c.1994; McNelis & Burke 2004; McNelis 2006)
  - subsidies (Hills 1988; Malpass 1990)
  - housing affordability (Milligan 2003; Yates et al. 2007)
  - social cohesion (Hulse & Stone 2007)
  - anti-social behaviour and stigma (Jacobs et al. 2003; Jacobs & Arthurson 2003; Atkinson & Jacobs 2008)
  - social mix (Atkinson 2008)
  - asset management (Gruis & Nieboer 2004; van Overmeeren & Gruis 2011; Kenley et al. 2009; Kenley et al. 2010)
  - the meaning of home (Clapham 2005; King 2008; Blunt & Dowling 2006)
  - housing discourses (Dodson 2007; Arthurson & Jacobs 2009)
  - governance, legislation, regulation, management and organisational structure (Mullins 2006; Walker 2000; Bradley 2008; Travers et al. 2011; Industry Commission 1993a, 1993b)
  - future prospects (Burke & Zakharov 2005; Jacobs et al. 2010a, 2010b; Maclennan & More 1997; Hills 2007).
- Works with different purposes such as:
  - theoretical research (Ball 1986, 1998; Ball and Harloe 1992)
  - practical research such as that undertaken under the auspices of AHURI, the Joseph Rowntree Foundation etc.
  - government inquiries and reports (UK House of Commons Communities and Local Government Committee 2012; National Housing Strategy 1992)
  - policy development and its implementation by housing organisations
  - policy analysis and advocacy by community, tenant and industry organisations (Shelter England 2013; National Shelter 2009; ACOSS 2008; Tenants Union of Victoria 2010; Planning Institute Australia 2007; Housing Industry Association 2008–9).

Thus, housing research is quite diverse. Such diversity is highly valued among researchers with different perspectives raising challenging and fundamental questions. However, it is also very fragmented, with housing researchers having little sense of how different types of research relate to one another.[1] Each researcher presents a different perspective on housing. Each has something significant to say about housing. Each, either implicitly or explicitly, would hope or even expect that what they had to say would have some impact on housing, either directly by changing some decision or indirectly by changing our understanding or our appreciation of housing.

The problem faced by decision-makers is whether and how to incorporate this disparate array of new meanings into their decisions. Of course, one strategy is simply to ignore or dismiss some or all of these new meanings – after all, it does complicate their decisions. Another strategy is to take on board those that accord with their own perspective or concerns. Even if they were open to these various new meanings, they still face some difficult prior questions. Each author purports to make some progress, some discovery or rediscovery of something important. How then do decision-makers get 'up to speed' with ever new research and new policy proposals? How are they to assess whether it is relevant to their situation? How do these new meanings relate to one another? How do the remote meanings of philosophy, ontology and epistemology or of theory relate to the day-to-day practicalities of decision-making and its implementation in new practices?

It is not just decision-makers who face difficult and complex issues, however. Just as progress in housing is not easy, progress in housing research is also not easy. Housing researchers suffer the same limitations and constraints as do decision-makers. They too make their own decisions as to their vision for housing and it is within that context that they decide what is important to investigate, what data and what theoretical framework is relevant, and what policies and directions will better realise that vision. Just as decision-makers bring with them their blind spots, their history and its economic, social, political and cultural biases and prejudices, so too do researchers. Such biases and prejudices may support those of the decision-makers and so we have situations similar to that documented by Chris Allen (2009) where researchers impose their cultural expectations of one group upon another and then develop methods of data collection and conceptual frameworks which justify these expectations.

Housing research seeks to address the problems confronting social housing. Yet, the history of actual research is economically, socially, politically and culturally biased in order to preserve and/or extend the interests of a particular group. It does not squarely address the problems. Is there any way out of this impasse?

If housing researchers are to provide decision-makers with practical advice, we need to envisage some method for addressing this complex of problems, questions and issues.

Housing researchers are increasingly recognising the need for a framework for interdisciplinary or transdisciplinary collaboration (Maclennan & Bannister 1995; Winter & Seelig 2001; Sibley et al. 2003; Lawrence 2004; Lawrence & Després 2004; Bridge et al. 2011; Phibbs & Thompson 2011). *Housing, Theory and Society* describes itself as a journal that 'furthers the agenda of housing research as an integrated, multidisciplinary field that is theoretically informed and embedded in wider societal issues' (Housing Theory and Society 2013). While the aspiration is strong, the difficulties and obstacles of working across disciplines are very apparent. There is a real need to find a common framework within which many different disciplines can operate.

Notes

1   In a broader context, others have also commented on the fragmentation of science. For instance, it was on ongoing concern of the quantum physicist David Bohm. See in particular Chapter 1 'Fragmentation and wholeness' in his book *Wholeness and the Implicate Order* (Bohm 2002). Edmund Husserl also criticises this fragmentation of science into endless specialties – see Lonergan's analysis of Husserl's *The Crisis of European Sciences and Transcendental Phenomenology* (Lonergan 2001: Chapter 11).

2 Functional Collaboration: Background

> Method is not a set of rules to be followed meticulously by a dolt. It is a framework for collaborative creativity. It would outline the various clusters of operations to be performed by [researchers] when they go about their various tasks. A contemporary method would conceive those tasks in the context of modern science, modern scholarship, modern philosophy, of historicity, collective practicality and coresponsibility.
>
> Bernard Lonergan in *Method in Theology* (1972 [1990]: xi)

Functional Collaboration, or functional specialisation, was a discovery of Bernard Lonergan (1904–1984), a Canadian methodologist, philosopher, theologian and economist.

Philip McShane, a long-time proponent of Functional Collaboration, has likened its discovery to that of the periodic table by Mendeleev in 1869, a paradigm shift that eventually transformed chemistry in the nineteenth century (McShane 2007b, ch.7: 7). The discovery of Functional Collaboration is, however, a discovery of different order. It is the discovery of the process that brings about progress and scientific discoveries.

So, I begin this chapter by briefly considering the remote orgins of Functional Collaboration in the work of Lonergan. I say 'remote origins' to distinguish Lonergan's discovery and expression of that discovery from its proximate origins within each researcher.

Initially, the writings of Mendeleev and Meyer on the periodic table attracted little interest. Similarly, the significance of Functional Collaboration has taken some time to be understood and appreciated, even among theologians, the audience at which it was first directed. While it emerged in a theological context, its relevance extends beyond theology.

As McShane suggests, however, Functional Collaboration is not simply the discovery of one person but rather, more deeply, the emergence of a 'serious group of global inquirers' (2004d: 2) seeking to meet the demands of history for a new way forward: 'The division of labour is suggested by the fermentation of centuries' (2002: 66). 'It is a division of labour based on needs emergent within any discipline' (2008: 5). So, scholars in different disciplines (including housing research) are seeking or proposing a more scientific approach to their discipline in order to meet the complex of issues arising within that discipline. These approaches, in some respects, parallel Lonergan's discovery of Functional Collaboration.

Bernard Lonergan and the remote origins of Functional Collaboration

Bernard Lonergan first published his discovery of Functional Collaboration as a journal article 'Functional Specialties in Theology' in *Gregorianum* (Lonergan 1969). This article, along with an extended introduction and further elaboration of the functional specialties, was published in 1972 as Chapter 5 of *Method in Theology* (Lonergan [1972] 1990).[1]

It is extremely difficult, and apt to be very misleading, to summarise the major directions of a thinker such as Bernard Lonergan. The following are some very general pointers to indicate the territory within which he operated and out of which he discovered Functional Collaboration. More precise meanings of what he was up to can only be had by a slow and thorough reading of his writings.[2] Here I will discuss Lonergan's search for a practical theory of history and his writings on functional specialisation.

Lonergan's search for a practical theory of history

Lonergan explored the deep desires of the human heart: to love and to be loved; to live in solidarity with one another; to express and create; to discover meaning in our lives and to understand; to grow and to improve ([1957] 1992; [1972] 1990; see also Moore 1989). Yet, despite these deep desires, he was confronted, as we are too, by a human community in a deep mess: the distortion, the oversight, the 'dumbing-down' and even the destruction of these deep desires; the fragmentation of a world divided and exploited by the ideologies of racism, classism, nationalism, colonialism and sexism; material poverty in rich countries and even greater poverty, famine and destitution in other countries; the periodic failure of economies; corrupt and failing political systems; the emergence of bureaucracy and state control; cultures which fail to provide satisfactory answers to our questions about living; the failure of education to cultivate and develop attentive, intelligent, reasonable and responsible persons; personal histories of betrayal, repressed dreams and minimal expectations; the 'existential' gap between the self we are and what we need to learn and become to respond adequately to new situations.

On the one hand, he discerned and articulated the deep desires at the core of our humanity. On the other hand, he was deeply aware of the human situation, what we have created historically. In some measure, we express our desires, we operate and co-operate in meeting these desires: we have created a complex infrastructure of tasks, roles and institutions, ways of collaborating with one another to meet our vital, social, cultural and personal needs and desires. Throughout history, we have slowly worked out better ways of providing food and shelter, education and justice, health and recreation, arts and leisure; better ways of relating to one another; better ways of discovering meaning. Rather than achieving these things alone, we have worked out how they can be done collaboratively with different elements being shared amongst many people; we have worked out how we can make improvements in each of the elements.

In some measure, however, we have not met the challenge of ongoing improvement in our institutions, in our ways of collaborating to meet these core human desires. So, for example, economic exploitation has had its justifications in racism, classism, nationalism, colonialism and sexism where individuals and groups aggrandise themselves at the expense of others:

> How, indeed, is a mind to become conscious of its own bias when that bias springs from a communal flight from understanding and is supported by the whole texture of a civilization? How can new strength and vigor be imparted to the detached and disinterested desire to understand without the reinforcement acting as an added bias? How can human intelligence hope to deal with the unintelligible yet objective situations which the flight from understanding creates and expands and sustains?
>
> (Lonergan [1957] 1992: 8–9)

The challenge of our deep desires pushes us to look forward to something better, something beyond the narrow confines of our individual and group biases: 'The challenge of history is for man progressively to restrict the realm of chance or fate or destiny and progressively to enlarge the realm of conscious grasp and deliberate choice' (Lonergan [1957] 1992: 253).

The pressing question, then, which Lonergan sought to address is: how do we achieve progress – a moving forward in an integral and concrete way? This question is not new. It is one raised by Plato over 2,000 years ago in *The Republic* ([360 BC] 1892). His answer was to make the philosopher king. Nevertheless, he also recognised the inadequacy of this answer. It is a question which many major philosophers have sought to answer (Bury 1955).

How are we to move forward? What are we to do? Indeed, what is the problem that we need to confront? Lonergan was not interested in some grand theory of history but rather in a practical theory of history (O'Leary 1998), one which looked to the future:

> So far from granting common sense a hegemony in practical affairs, the foregoing analysis leads to the strange conclusion that common sense has to aim at being subordinated to a human science that is concerned, to adapt a phrase from Marx, not only with knowing history but also with directing it.
>
> (Lonergan [1957] 1992: 253)[3]

Lonergan's solution to this problem had a long gestation. It began with a retrieval of the 'intellectualist, dynamic, and existential' metaphysics of Thomas Aquinas (and Aristotle) (Crowe 1973; Lawrence 1978; Lonergan 1997, 2000). This metaphysics was in stark contrast to a decadent scholasticism and to the conceptualist and essentialist metaphysics of Scotus and Suarez etc. which have been so trenchantly criticised by recent philosophers such as Heidegger (Lawrence 1978: 53).

In his seminal work *Insight: A Study of Human Understanding* (Lonergan [1957] 1992),[4] Lonergan explored the dynamics of understanding. His starting point is that every day creative moment of 'insight', a phenomenon largely overlooked in the history of philosophy. His goal, as he puts it, is this:

> Thoroughly understand what it is to understand, and not only will you understand the broad lines of all there is to be understood but also you will possess a fixed base, and invariant pattern, opening upon all further developments of understanding.
>
> (Lonergan [1957] 1992: 22)

Despite the apparent remoteness of this goal, *Insight* had a practical goal. In his preface he asks: 'What practical good can come of this book?' He answers as follows:

> The answer is more forthright than might be expected, for insight is the source not only of theoretical knowledge but also of all its practical applications, and indeed of all intelligent activity. Insight into insight, then, will reveal what activity is intelligent, and insight into oversights will reveal what activity is unintelligent. But to be practical is to do the intelligent thing, and to be unpractical is to keep blundering about. It follows that insight into both insight and oversight is the very key to practicality.
>
> Thus, insight into insight brings to light the cumulative process of progress. For concrete situations give rise to insights which issue into policies and courses of action. Action transforms the existing situation to give rise to further insights, better policies, more effective courses of action. It follows that if insight occurs, it keeps recurring; and at each recurrence knowledge develops, action increases its scope, and situations improve.
>
> Similarly, insight into oversight reveals the cumulative process of decline. For the flight from understanding blocks the insights that concrete situations demand. There follow unintelligent policies and inept courses of action. The situation deteriorates to demand still further insights, and as they are blocked, policies become more unintelligent and action more inept. What is worse, the deteriorating situation seems to provide the uncritical, biased mind with factual evidence in which the bias is claimed to be verified. So in ever increasing measure intelligence comes to be regarded as irrelevant to practical living. Human activity settles down to a decadent routine, and initiative becomes the privilege of violence.
>
> (Lonergan [1957] 1992: 7–8)

In *Insight*, Lonergan did not reach a full solution to the concrete problem of progress in human history. Rather, he more clearly defined the problem as he explored the role of 'insight' in mathematics, science and common sense, identified the flights from understanding that block insight, and redefined metaphysics as the explication of the implicit grounds on which we operate and as the 'integral heuristic structure' of our knowing and doing. In that 'personal appropriation of the concrete dynamic structure immanent and recurrently operative in his own cognitional activities' ([1957] 1992: 11), Lonergan identified the grounds for the integration of all forms of knowledge and interdisciplinarity:

> If to convince oneself that knowing is understanding, one ascertains that knowing mathematics is understanding and knowing science is understanding and the knowledge of common sense is understanding, one ends up not only with a detailed account of understanding but also with a plan of what there is to be known. The many sciences lose their isolation from one another; the chasm between science and common sense is bridged; the structure of the universe proportionate to man's intellect is revealed; and as that revealed structure provides an object for a metaphysics, so the initial self-criticism provides a method for explaining how metaphysical and antimetaphysical affirmations arise, for selecting those that are correct, and for eliminating those that patently spring from a lack of accurate self-knowledge. Further, as a metaphysics is derived from the known structure of one's knowing, so an ethics results from knowledge of the compound structure of one's knowing and doing; and as the metaphysics, so too the ethics prolongs the initial self-criticism into an explanation of the origin of all ethical positions and into a criterion for passing judgment on each of them.
>
> (Lonergan [1957] 1992: 23)

Further, he was able to relate common sense and science:

> Perhaps it now is evident that the whole of science, with logic thrown in, is a development of intelligence that is complementary to the development named common sense. Rational choice is not between science and common sense; it is a choice of both, of science to master the universal, and of common sense to deal with the particular.
>
> (Lonergan [1957] 1992: 203)

While *Insight* does not reach a full solution to the problem of human living, it does reject some solutions and point to the elements of that full solution in what he called 'cosmopolis'. This 'is concerned with the fundamental issue of the historical process. Its business is to prevent practicality from being shortsightedly practical and so destroying itself' (pp.263–264). It is 'concerned to make operative the timely and fruitful ideas that otherwise are inoperative' (p.264). It is not a 'police force' (p.263) by which one group imposes itself and its ideas on another. It is not 'a busybody' (p.264). 'It does not waste its time and energy condemning individual egoism...It is not excited by group egoism' which, in time, 'generates its own reversal' (p.264). 'It is not a group denouncing other groups; it is not a super-state ruling states; it is not an organization that enrols members, nor an academy that endorses opinions, nor a court that administers a legal code' (p.266). 'It is not easy. It is not a dissemination of sweetness and light, where sweetness means sweet to me, and light means light to me' (p.266). Cosmopolis 'protect[s] the future against the rationalization of abuses and the creation of myths' (p.265). 'It is a withdrawal from practicality to save practicality. It is a dimension of consciousness, a heightened grasp of historical origins, a discovery of historical responsibilities' (p.266). '[T]wo allies can be acknowledged. On the one hand, there is common sense, and in its judgments...common sense tends to be profoundly sane. On the other hand, there is dialectical analysis; the refusal of insight betrays itself; the Babel of our day is the cumulative product of a series of refusals to understand; and dialectical analysis can discover and expose both the series of past refusals and the tactics of contemporary resistance to enlightenment' (p.267).

In the last chapter of *Insight*, Lonergan returns to this problem of progress in history and develops an extended heuristic structure of the solution, identifying thirty-one elements of that solution. It is not the place here to outline this heuristic structure (and to make sense of it requires a close reading of the previous chapters). However, one element of this solution is collaboration and its grounds in 'belief'. As Lonergan notes in *Method*: 'To appropriate one's social, cultural, religious heritage is largely a matter of belief' ([1972] 1990: 41).[5] He goes on to note the significant role that belief plays in the advancement of science: rather than continually repeating someone else's work, the scientist tends to assume the results of previous work and builds upon them. It is when this work runs into problems that previous results are scrutinised more closely. Results are verified and falsified as they are incorporated into new work: 'Human knowledge, then, is not some individual possession but rather a common fund, from which each may draw by believing, to which each may contribute in the measure that he performs his cognitional operations properly and reports their results accurately' (p.43). An extended excursus in the final chapter of *Insight* provides the grounds for belief and for human collaboration in meeting the problem of progress in history. However, the unanswered question was: how are we to divide up the work such that the work of one person could draw upon as well as contribute to the work of others?

Lonergan's discovery in 1965 of functional specialisation as 'a framework for collaborative creativity' was his long sought after solution to the problem of progress in history.

Functional specialisation

As Lonergan concluded in *Insight*, the problem of progress is not simply a problem of understanding but also a problem of implementation, of finding a way forward. Further, it is not simply a problem of understanding and implementation once and for all, but of developing understanding and ongoing implementation. It is, then, a problem of method, but one which links developing understanding and continually implementing new solutions in an ongoing cyclic way.

A preliminary notion of method

> is a normative pattern of recurrent and related operations yielding cumulative and progressive results. There is a method, then, where there are distinct operations, where each operation is related to the others, where the set of relations forms a pattern, where the pattern is described as the right way of doing a job, where operations in accord with the pattern may be repeated indefinitely, and where the fruits of such repetition are, not repetitious, but cumulative and progressive.
>
> (Lonergan [1972] 1990: 4)

The empirical method of science is one set of recurrent and related operations that produces results. By linking this set of operations with the scientist as operating subject, Lonergan extends or generalises this empirical method:

> Generalized empirical method operates on a combination of both the data of sense and the data of consciousness: it does not treat of objects without taking into account the corresponding operations of the subject: it does not treat of the subject's operations without taking into account the corresponding objects.
>
> As generalized empirical method generalizes the notion of data to include the data of consciousness, so too it generalizes the notion of method. It wants to go behind the diversity that separates the experimental method of the natural sciences and the quite diverse procedures of hermeneutics and of history. It would discover their common core and thereby prepare the way for their harmonious combination in human studies.
>
> (Lonergan [1974] 1985: 141)

As explored in *Insight*, this common core of operations – experiencing, understanding, judging and deciding – underpin both common sense and scientific method. Within science, this common core of operations is utilised in different ways for different ends. So, in collecting data, the operations are grouped in one way; in seeking to understand this data, they are grouped in another way; in affirming what is (and distinguishing this from what might possibly be), they are grouped in another; in acting upon the data, they are grouped in another (Lonergan [1972] 1990: 6–13). This correspondence between objects and the operations of the subject provides the grounds on which Lonergan distinguishes between the functional specialties. So, in *Method* he outlines his discovery that progress in human living is achieved in collaboration with others through the totality of eight functional specialties which are internally related to one another.

When Lonergan wrote *Method,* theology, as with many other human sciences, was fragmented with many diverse specialisations developing such that different people doing differing things had little understanding of how their work related to other pieces of work or how their work related to progress in theology. Specialties had developed in two ways. The first way, field specialisation, divides and sub-divides the field of data. For example, history gets divided into different periods, into different countries, into different groups and then into groups within periods within countries. The second way, subject specialisation, divides and sub-divides by classifying the results of investigations. For example, human sciences are divided into subjects such as philosophy, psychology, economics, politics, cultural studies etc. (Lonergan [1972] 1990: 125–126).

A third way of developing specialties is functional specialisation which 'distinguishes and separates successive stages in the process from data to results' (Lonergan [1972] 1990: 126). In this way,

> functional specialties are intrinsically related to one another. They are successive parts of one and the same process…are functionally inter-dependent…without prejudice to unity, it divides and clarifies the process from data to results…it provides an orderly link between field specialization… and subject specialization.

In *Method* ([1972] 1990: 127–133), Lonergan is specifically outlining a method for progress in theology. He distinguishes eight functional specialties as follows:

Understanding the Past

1   Research 'makes available the data relevant to theological investigation' (p.127).
2   Interpretation 'understands what was meant' in this data, 'in its proper historical context, in accord with its proper mode and level of thought and expression, in the light of the circumstances and intention of the writer' (p.127).
3   History 'grasps what was going forward in particular groups at particular times and places' (p.178).
4   Just as 'empirical science aims at complete explanation of all phenomena, so Dialectic aims at a comprehensive viewpoint' (p.129). It is concerned with conflicts within historical movements and with diverging viewpoints. It seeks a viewpoint from which it can understand these conflicts and diverging viewpoints. It brings to light differences that are irreducible and differences that are complementary.

Looking to the Future

5   Foundations provide a basic orientation and, thus, the objectification of the fundamental existential horizons of human knowing and doing as well as the transformations of the subject and their world as both historical and communal.
6   Doctrines express judgements of fact and judgements of value and so are concerned with affirmations and negations about what is known and how to live.

7   Systematics 'works out appropriate systems of conceptualization' (p.132), removes apparent inconsistencies and promotes coherence.

8   Communications is concerned with external relations, with communicating results to other disciplines, with transposing results into other cultures.

These eight specialties operate in two phases, the first four specialties in the first phase and the second four specialties in the second phase:

> In the first phase one begins from the data and moves through meanings and facts towards personal encounter. In the second phase one begins from reflection on authentic conversion,[6] employs it as a horizon within which doctrines are to be apprehended and an understanding of their content sought, and finally moves to a creative exploration of communications differentiated according to media, according to classes of men, and according to common cultural interests.

> (Lonergan [1972] 1990: 135–136)

The need for some division of labour is clear from the way work has been increasingly divided according to field and subject. Why does Lonergan, however, argue for a division of labour according to functional specialties? First, unlike field and subject specialisation which 'divide the same sort of task among many hands', functional specialisation distinguishes tasks and prevents them from being confused: 'Different ends are pursued by employing different means, different means are used in different manners, different manners are ruled by different methodical precepts' (pp.136–137). Second, without distinctions between particular ends, the tasks to be performed to achieve these ends and their associated methodical precepts, 'investigators will not have clear and distinct ideas about what precisely they are doing, how their operations are related to their immediate ends, and how such immediate ends are related to the total end of the subject of their inquiry' (p.137). Third, functional specialisation curbs totalitarian ambitions. One type of question cannot dominate another and, while we may do our work alone, we depend upon and build upon the work of others. Fourth, functional specialties 'resist excessive demands'. Each makes a specific contribution to a complex range of issues and produces 'the type of evidence proper to the specialty'.

The discovery of functional specialties is one of the crowning achievements of Bernard Lonergan's life. It was achieved late in his life just before he became very ill.[7] In many ways, *Method* is a very different book from *Insight*. Most obviously is the shift in expression from the faculty psychology of Thomas Aquinas to the intentionality analysis of phenomenology.[8] More significantly, Lonergan's strategy in his earlier magnum opus *Insight* is missing from *Method*. In *Insight* the reader is challenged by his relentless pedagogical drive in which he poses a set of questions and in answering them moves on to pose a new set. *Method,* however, is more compact. Moreover, Lonergan not only has the task of outlining his discovery of the functional specialties, but also the task of addressing key methodological problems of contemporary theology. As a result, *Method* intersperses an exposition of the functional specialties with a discussion of other problems in theology and does not treat the functional specialties in a thoroughly systematic way. The functional specialties are not well developed leaving readers with the difficult task of 'unpacking' his meaning and working out their significance for themselves. This poses some problems for developing a thorough understanding of functional specialisation and for its application in different disciplines.

Searchings for Functional Collaboration

Lonergan developed his discovery of functional specialisation within the context of theology. In this context, functional specialisation constitutes progress in theology and continually 'mediates between a cultural matrix and the significance and role of a religion in that matrix' (Lonergan [1972] 1990: xi). The canvas on which Lonergan was operating, however, was the totality of world process and, within that, the totality of human history. He was searching not just for a method in theology but rather for a practical theory of history.

While his discovery was first put forward in the context of theology, its wider significance was noted (as a criticism) by another prominent theologian, Karl Rahner: 'Lonergan's theological methodology seems to me to be so generic that it actually suits every science' (Rahner 1971, quoted in McShane 2009: 23 and fn 120). In this larger context, functional specialisation has relevance beyond theology.

The emergence of Functional Collaboration can be discerned in disciplines as diverse as literature, deep ecology and comparative housing research where authors have had no exposure to the discovery of Functional Collaboration but have spontaneously sought to structure their work around some of the functional specialties. Of particular interest to housing scholars is the work of Michael Oxley in comparative housing research. In other disciplines such as feminist studies, language studies, economics, and the natural and formal sciences, authors who have affirmed the importance of Functional Collaboration have illustrated the possibilities of structuring these disciplines in the light of this discovery.

Theory of Literature by Rene Wellek and Austin Warren

In their preface to the first edition of *Theory of Literature*, Rene Wellek[9] and Austin Warren[10] (1963: 8) note how their book differs from previous books on literature and describe it as follows: 'We have judged it of central use to ourselves and others to be international in our scholarship, to ask the right questions, to provide an organon of method.'

In this 'organon of method', they are primarily concerned with understanding the past. In relation to functional specialties, what is significant about their book is not so much its content, but the way in which the headings of chapters (and some of the discussion) point to and seek to distinguish different types of work within literature studies.

Chapter 1 distinguishes between literature as a creative art and literary study as a species of knowledge. Wellek and Warren regard these as two distinct activities (p.15), thus, distinguishing between the world of common sense (literature as a creative art) and the world of science (literary study).

Chapter 6 discusses issues in relation to 'the ordering and establishing of evidence': 'One of the first tasks of scholarship is the assembly of its materials, the careful undoing of the effects of time, the examination as to authorship, authenticity, and date' (p.57). Or again: 'Among these preliminary labours one has to distinguish two levels of operations: (1) the assembling and preparing of a text; and (2) the problems of chronology, authenticity, authorship, collaboration, revision...'

Chapter 2 begins with some questions about the subject matter of literary scholarship: 'What is literature? What is not literature? What is the nature of literature?' (p.20). This is a question of Interpretation, a question which requires answering before any particular piece of literary scholarship is interpreted.

Chapter 4 distinguishes theory, criticism and history:

> Within our 'proper study', the distinctions between literary theory, criticism, and history are clearly the most important. There is, first, the distinction between a view of literature as a simultaneous order and a view of literature which sees it primarily as a series of works arranged in a chronological order and as integral parts of the historical process...Of course, 'literary criticism' is frequently used in such a way as to include all literary theory; but such usage ignores a useful distinction. Aristotle was a theorist; Sainte-Beuve, primarily a critic. Kenneth Burke is largely a literary theorist, while R.P. Blackmur is a literary critic. (p.39)[11]

Wellek and Warren then go on to describe these distinctions between theory, history and criticism as 'fairly obvious and rather widely accepted'. They continue: 'less common is the realization that the methods so designated cannot be used in isolation, that they implicate each other so thoroughly as to make inconceivable literary theory without criticism or history, or criticism without theory and history, or history without theory and criticism'.

Wellek and Warren are seeking to bring some order to literary studies by distinguishing between 'ordering and establishing evidence', theory, criticism and history. This parallels the first four functional specialties of Functional Collaboration: Research, Interpretation, History and Dialectic.

Arne Næss on implementing deep ecology

Arne Næss (1912–2009) was one of Norway's most eminent philosophers and founder of the 'deep ecology' movement. Næss was interested in keeping together a movement which was concerned about a new future relationship between humanity and nature. His orientation was moving forward, looking to the future rather than understanding the past.

As part of a special double issue of the *Ecologist* devoted to 'Rethinking Man and Nature: Towards an Ecological Worldview', Næss (1988) discussed the ultimate premises of deep ecology. He argued for a broad platform to the deep ecology movement that would draw together supporters from disparate backgrounds – different religions and philosophies. In conjunction with George Sessions, he formulated an eight-point platform (p.130). He then proceeded to relate the eight points to (i) higher level or ultimate premises by asking 'from which premises, if any, are the points derived?' and (ii) lower level views 'which have one or more of the eight points as part of their set of premises' (pp.130–131) and also lower level practical decisions. A practical decision is 'derived in part from details of a particular situation' and so 'will always involve premises in addition to those of the upper levels' (p.131).

Næss argued that there may be 'other and better proposals for a platform, but I expect that a distinction between the 4 levels will be of importance.' He goes on: '[e]ven moderately integrated people have reasons for their views – or, at least, they indirectly pretend to have reasons. What is unfamiliar, perhaps, is the relation between more philosophical or general views and the concrete, and also the demand for clear articulation of that relationship'.

He concluded the article by expressing the hope that it 'shows how Deep Ecology views can manifest both plurality and unity'.

Arne Næss sought to bring some order into the implementation of a new relationship between humanity and nature. He identified four levels in the process from ultimate premises to practical decisions, but tended to envisage the movement from ultimate premises to practical decision as a logical process (though he did acknowledge 'premises in addition to those of the upper levels' (p.131). This parallels the four implementation functional specialties of Functional Collaboration: Foundations, Policies, Systematics and Communications.

Comparative housing research

In the 1990s, debates took place about the possibility, purpose and methodologies appropriate for comparative housing research. In a series of articles, Michael Oxley (1989, 1991, 2001) developed his view of comparative housing research, culminating in a key article which examined its aims and methodologies. He concluded that 'one of the greatest confusions in housing research that covers several countries is to box all such work together and call it 'comparative'' (2001: 103). He further suggested that 'the use of the term "comparative housing research" should be limited to research that genuinely compares and contrasts'. He continued: 'the best of such research produces plausible, evidence-based conclusions about the reasons for the similarities and differences'. In consequence, he recognises how difficult 'high-level' comparative housing research is and how much work is required to do it successfully.

Oxley advocates 'a more scientific approach' (p.90) to comparative housing research and notes that '[i]t is likely that, to be successful, teams of nationally based researchers would have to collaborate on such projects' (p.103).

He proposes that we divide up the work between four teams: explorers, empiricists, theorists and scientists. In my view, explorers and empiricists have the same purpose and associated method, except that empiricists are dealing with material about which something is known and explorers are dealing with material about which little, if anything, is known (except to practitioners in a particular country). Given the current state of housing studies and his interest in high-level comparison, Oxley sets aside the work of explorers (who are generally interested in low-level comparison). He then goes on to propose that research teams be formed according to their particular purpose and method. Thus, empiricists 'find more facts and collate and organise these facts' (p.91); theorists 'provide ideas to make sense of facts and they may build models and formulate hypotheses' (p.92); scientists 'test hypotheses' (p.93).

Each of these teams has a different purpose:

> International housing research should not be driven by a single methodological approach. Methods and purpose should go together. A variety of aims demand a variety of approaches. Methodology has to be fit for the purpose and it should be explicit. Too much housing research is without a clear method that has been reasoned to be the best way of tackling a particular issue. This is a reflection of a lack of theorising and a concern to engage in lots of description. (p.103)

It is this linking of method and purpose that characterises his article, particularly in his exposition of explorers, empiricists, theorists and scientists.

After his review of different types of comparative analyses, Oxley moves on to the issue of the transferability of housing policy. He notes that while 'the purpose of a large number of international housing studies...is to improve policy and practice', they 'typically stop short of transferability analysis'. He further notes that 'there is a strong case for making transferability issues more explicit aspects of such work and developing analytical methods that are focussed on housing transferability' (pp.98–99). To this end he proposes an examination of policies in seven contexts using the example of the transfer of a form of points system rents (PSR) from country A to country B as follows:

i   Investigate the intended purpose of points system rents (PSR) in country A;
ii   Consider whether it did achieve its objectives in A;
iii Investigate the reasons for it achieving its objectives in A;
iv   Investigate the role of a wide set of circumstances including social, political, economic, and institutional factors in influencing the outcomes in A;
v   Examine the purpose of social housing rents in country B;
vi   Investigate the feasibility of introducing PSR in B. The roles of social, political, economic, and institutional factors in influencing potential feasibility could be examined;
vii Predict the consequences of introducing PSR in B. This would have to take into account forecasts about a wide range of housing and contextual matters. (p.98f)

While Oxley speaks of seven contexts of transferability (and shifts from speaking of purposes and teams to the process of transferability through each context), we can distinguish within these contexts four different purposes (each with some yet to be determined associated method).

The first four contexts (contexts i–iv) are concerned with an already operating housing policy, the points system in country A. Together they contribute to an evaluation of this policy. In this sense they build upon the work of empiricists, theorists and scientists.

The last three contexts (contexts v–vii) are concerned with whether and how this policy might operate in another country. They show the difficulty of implementation and of transferring housing policies and practices from one country to another. Oxley highlights the need to make this process more explicit. Again, these contexts go beyond the work of the research teams (empiricists, theorists and scientists) and also evaluation, the work of evaluators.

So, we could characterise context (v), 'examining the purpose of social housing rents in country B', as the work of 'policy-makers' for it concerns the formulation of some new direction in social housing rents in country B; we could characterise context (vi), 'investigating the feasibility of introducing', a new policy in the context of 'social, political, economic and institutional factors', as the work of 'strategic planners' for it concerns introducing something new into an already existing situation and this must be taken into account; we could characterise context (vii), predicting 'the consequences of introducing' a new policy, as the work of 'practitioners' for it concerns what actually happens on the ground.

While each of these contexts of transferability has a purpose, Oxley does not link them to his previous discussion on the research teams. Such a connection which more explicitly links policy with research would add support to his view that 'the value of international housing research in policy and practice circles would be greatly enhanced if rather more sophistication were to accompany suggestions about trying something from elsewhere' (p.99).

Here, the key element in Oxley's article is his search for a more scientific approach to comparative housing research and his proposal for some sort of division of labour among research teams and the transferability of housing policy. This proposal is not based on data or results, but rather upon stages in the process from data to results, with each stage distinguished according to some purpose and associated method.[12]

Alexandra Gillis-Drage on the third wave of feminism

Alexandra Gillis-Drage (Drage 2005; Gillis-Drage 2010) examined sixteen feminist journals[13] as well as another dozen journals that regularly include feminist content. As she was doing so, she was asking, as other feminists have done particularly in the third wave of feminism (Grosz 2000; Baumgardner & Richards 2003),[14] 'the number one question about feminism…feminism's *raison d'être*, its purpose, its reason for being' (Gillis-Drage 2010: 7) or, to put it another way,

'What can *you and I* do together, communally, to better understand and implement feminism's future?' (Gillis-Drage 2010: 7–8).

She notes the ferment within feminism as it seeks to come to terms with the shift from political action to reflective discourse (particularly within the academy) and also notes a distinct change in the second wave of feminism:

> Where the first wave was a concentrated practical effort toward political change (especially toward attaining 'the vote'), the second wave, though continuing the activist position, introduced the necessity of a new dimension in feminism: *reflection* on women, by women, to be expressed and communicated through books and newly established journals, and through the establishment of academic departments devoted to women's studies.
>
> (Gillis-Drage 2010: 10)[15]

Gillis-Drage identifies this change as a shift in method as feminists seek their particular goal of progress for women. While this goal is common to all feminisms, there are particular differences as 'different groups, races, cultures, ages, nationalities, religious beliefs, personal experiences, etc., of women around the world usher in differing sets of concerns and concrete circumstances' (Gillis-Drage 2010: 11). She notes Barbara Smith's widely inclusive definition, dating from the 1970s:

> feminism is the political theory and practise to free *all* women: women of color, working-class women, poor women, physically challenged women, lesbians, old women, as well as white economically privileged heterosexual women. Anything less than this is not feminism, but mere female aggrandizement.
>
> (Gillis-Drage 2010: fn 16)[16]

Gillis-Drage identifies the key problem as follows:

> Feminism's goal of progress for women is, by its very nature, both practical *and* theoretical. On the practical side, feminism's concern is for women's lives in their concrete functioning; for human living in all its concrete conditions, economic, ecological, cultural, religious, educational, political, and so on; and for effective practical change to those conditions. But in any situation, in any society, before practical change can occur, there is invariably the need for reflection. Preceding practical action, there is a *need to understand* the full situation: *What has gone on in the past? What is the present state of women's living? What needs to change in the future? How does it need to change?* In other words, this general activity of *understanding* is a prior condition for practical change.
>
> (Gillis-Drage 2010: 12)

What is missing is some unifying scheme which would 'hold together' the diversity of particular feminisms that were represented both in the journals and in political activities. It is at this point that she introduces Functional Collaboration or, as she called it in her book, *Thinking Woman* (Drage 2005), 'the Great Circle of Feminism' (p.143ff).[17] She puts it forward as a way of resolving 'questions about a growing opposition between theory and praxis' (p.153). She briefly indicates how the work of feminism could be divided up into those tasks which seek to 'make sense of the past' (History, Interpretation and Research), those which seek to 'make sense of the future' (Policies, Systematics and Communications), the task of making sense of the many differing positions on progress (Dialectic) and the task of making sense of the whole cycle of moving back into the past and forward into the future (Foundations) (Gillis-Drage 2010: 14–17).

Gillis-Drage describes 'the Great Circle of Feminism' as 'a functional, practical way – that can be globally implemented – to think theoretically about and to provide concretely for the needs of local individuals and groups' (Drage 2005: 153). She concludes *Thinking Woman* as follows:

> In examining feminism, we have found that there is an exceptional cultural diversity to be taken into account, along with a demand for concrete thinking that will meet that diversity. The diversity that exists in feminism is indeed a problem-in-need-of-a-solution. But amidst this diversity there is to be discovered the deeper and more profound thinking, the questing, searching Women who are striving to bring about progress for themselves and the human group. The discovery of this heuristic openness, then, is radically a discovery of *method – our way of going forward in the feminist enterprise.* (p.154)

Shaping the Future of Language Studies by John Benton

In *Shaping the Future of Language Studies,* John Benton (2008) presents a methodological challenge to students and professionals struggling with the problem of linguistic universals and with the problem of integrating the broad field of language studies. He picks up the methodological effort of the Greenberg School of linguistics and looks forward to a methodological restructuring of language studies (p.1).

His starting point is the need for self attention among linguists, 'a serious self-attentive puzzling over the symbol '?'' (pp.2, 78), through which is discovered the 'core invariant patterns of quest' that bubble up spontaneously in language users and ground language universals. These 'core-attitudes' (what-attitudes, why-attitudes, is-attitudes, what-to-do-attitudes and is-to-do-attitudes) 'are elemental to the language user's self knowledge' (p.22). Using examples from English literature (Shakespeare, Joyce and Donne), *The Bhagavad-Gita* from Vedic literature and the story of Helen Keller, he unfolds a 'precise phenomenology of human consciousness', empirically verifying these 'core-attitudes' as the grounds for and expression in language universals. He then proceeds to show how these language universals 'ground generic linguistic performance' across different branches of the linguistic tree: Indo-European (Ancient Sanskrit and Modern Hindi), Semitic (Arabic), African (Swahili), Asian (Cantonese), Nordic (Swedish) and Teutonic (German).

Benton identifies two root causes for the failure of contemporary linguistics to appreciate the significance of these 'core-attitudes' in reaching adequate explanations of language universals and linguistic performance, viz. conceptualism and common-sense eclecticism, and critically evaluates the work of some prominent linguists, noting 'the culture of extra-scientific opinion'.[18] He then 'introduces the essential features of the two principles necessary for successful restructuring in language studies: first, a focal shift in grammatology and secondly, a functional relating of sub-fields of language' (p.99). The first principle, generalized empirical method, 'is the procedure of critical thinking that occurs when we are luminous, clear, about what we are doing when we are thinking' (p.100). This focal shift exposes 'a regrettable and systematic oversight in the writing of a group of very influential thinkers who have sought to achieve progress in the search for language universals' (pp.4, 80). The second principle, functional specialisation, is a 'way of organizing the entire enterprise of language studies' (p.104).

After a brief outline of the eight functional specialties that methodologically restructure language studies, Benton goes on to do a 'focused survey of ten major journals representing the spectrum of scholarship in the field of language studies' (p.114), noting the diversity and fragmentation of work within the journals:[19]

> For all the richness and diversity of data within the medium, the stakeholders lack a practical mandate grounded by a strategy committed to addressing the need for an efficient globally collective procedure within which to structure that richness and diversity in a manner that would yield cumulative and progressive results. (p.117)

Benton then undertakes some exercises in methodological heuristics. Beginning with the June 2001 issue of the *Cambridge Quarterly*, he illustrates how 'there emerge the 'shadows' of a number of functional specialties, randomly placed, not only within the journal listings, but also within various parts of the articles themselves' (p.119). In a second exercise, he carefully analyses three paragraphs from one article, 'Englands of the Mind' by David Gervais (2001), and shows the methodological shifting even within paragraphs. A third and fourth exercise analyse Wellek and Warren's *Theory of Literature* (1963)[20] and the Greenberg School as outlined in *Universals of Human Language* (Greenberg 1978). He shows how both literary studies and linguistic studies anticipate functional specialisation: 'even though a principle of integration is absent, there is a spontaneous struggle for a structure of reflection that vaguely anticipates a principle of integration' (Benton 2008: 4).

Benton concludes with some reflections on how generalised empirical method and functional specialisation would transform 'the future possibilities for the humanities in education'. He thus looks forward to a transformation of linguistics and language studies. He regards his book as incomplete: 'it should be noted that this work, though rounded off, is incomplete. It represents a firm halfway station to a full heuristics of basic linguistics and basic grammar' (2008: 5).

Pastkeynes Pastmodern Economics: A Fresh Pragmatism by Philip McShane

Philip McShane has written extensively on both Lonergan's discovery of functional specialisation (2005a, 2007a, 2007b) and Lonergan's economics (McShane 1980, 1998, 2002, 2007c; Anderson & McShane 2002; Lonergan [1942–1944] 1998, [1983] 1999).

Pastkeynes Pastmodern Economics: A Fresh Pragmatism (McShane 2002) is a complex book that interweaves reflections and pointers on Lonergan's later work on functional specialisation, his early work on economics (Lonergan [1942–1944] 1998) and on current and past economics. What results is a devastating critique of current economics and economic education.

His aim, in relation to economics, is 'to draw attention to the need for a new generalization in economics, a new theory, a new orientation' (2002: 36). The division of labour (i.e. functional specialisation) within economics is 'the key to the full, distant, transition to a democratic economics' (p.7). He regards Lonergan's economics as:

> a Copernican revolution in economic methodology. It is a double achievement: there is the discovery and precision of the significant variables; there is the sublation of Schumpeter's view of theory's dependence on the past in functional specialization…The former [the discovery of significant variables] strikes me not merely as a paradigm shift in a science but as a Galilean leap to science. The latter invention [functional specialisation] is a radical shift in the meaning of implementation within a New Order of *Die Wendung zur Idee*[21]…There are further aspects of this achievement that mesh with… a transformation that is not just Kondratiev-like but axial, a *pastkeynesian* theoretic not of employment but of leisure…
>
> (McShane 2002: 37)

McShane asks us to undertake 'a thought experiment' to 'find some order in the present spread of interests and printings' in order to find a pattern within economic studies (p.54ff). He suggests that this experiment will reveal two halves to economic studies: a focus on the past and a focus on the future. It will reveal a pattern of policy, planning and execution in some order: 'without planning there can be no executive reflection; without policy, the plan is not grounded. But what grounds the policy?' (p.54). In seeking an answer to that question, he suggests first of all that policy is grounded in some history. Secondly he notes 'the swing' from history towards the future, a swing 'represented by a vague collection of publications regarding *critical assessment* and *selection*' (p.55): 'So I come to suggest *A Foundation* in a broad loose sense: we are using the data of economic life to generate economic history and our economic future' (p.57). The grounds of policy, planning and executive reflection 'somehow lurk in the given of previous efforts'. They are the result of economic research, interpretation and history. Just as beyond policy there is an issue of grounds, so too beyond history there is a similar problem of grounds, 'a problem of a plurality of both origins and goals'. It gives rise to the differences in economic theories and requires some sorting out. It is a zone of work that McShane calls 'discernment'. He is at pains to point out that he is

> not inventing distinctions; we are ordering distinctions invented by centuries of searching in economics, richly evident now in journal articles that collect and organize economic data, that seek to interpret this economist or that, that grapple with the problem of putting in sequence not only ideas of the past but the events of the past. (p.57)

Pastkeynes Pastmodern Economics is an invitation to rethink current economics. It asks us to think about two different approaches to the IS–LM model[22] as well as an issue of *Economica*[23] devoted to theories of economic growth. McShane challenges us to discern what is going on in different articles, to work out the functional specialties in which the author is operating and to notice the shift (often rapid) from one to another (p.67). In the Appendix, McShane seeks to transpose into a new context the quantity theory of money which, according to Mark Blaug (1995), is the oldest surviving theory of economics. The central question (indeed, the central challenge of the new economics) is: *What Is Money?* (McShane 2002: 137ff).

In a chapter concerned with Lonergan's interlocutors in economics – Keynes, Marx, Marshall, Kalecki and Schumpeter, among others – McShane argues that '[j]ust as the eight *hodic*[24] zones lifts scholarly work out of randomness, so pragmatic advertence to the possibilities of dialogue permits specialists to know precisely whom to turn to for relevant enlightenment, who they are turning to in dialogue' (p.71).

Formal and natural sciences

Other authors explore the emergence of the functional specialties in the natural and formal sciences: McShane (1976, 2002, 2007a, 2007b) in physics, chemistry, biology and geometry and Terrance Quinn (2005) in mathematics. David Oyler (2010) and Terrance Quinn (2012) generalise the functional specialties from theology to the formal and natural sciences.

For instance, McShane (2002: 60–61) notes the emergence of the evident need for functional specialisation in a range of natural and formal sciences:

> Even a swift perusal of Biological Abstracts would reveal not only the existence of the four types of tasks in the areas of zoology and botany, but also the existence of large problems of discernment and canon-development *[Foundations]* regarding issues of classification, evolution, ontogenetics. The

science of chemistry, apparently so secure in its methods and identity as it emerged from the broad discoveries of Meyer & Mendeleev after the 1860s blossomed into a range of specializations, now stands in need of the suggested structure not only to give it collaborative unity but to make possible a discernment precisely of its identity in the face of the developments of physics.

Finally, there are the needs of physics itself. Even if one focuses only on the particular zone of particle physics, the twentieth century has thrown up a massive panoply of theoretical & practical problems, as well as problems of communication. The central issue in that field is the nature of real geometry, an issue which illuminates peculiarly the present cultural crisis.

McShane then goes on to an analysis of 'The Origins of Geometry', an appendix to Husserl's *The Crisis of European Sciences* (1970). The task McShane set himself is 'the problem of discerning, within the confines of a single article in any field, how much the author wanders around in the eight zones [of Functional Collaboration]' (2002: 61).

* * *

Scholars in various fields are seeking a more scientific approach to their discipline. While some have no knowledge of Functional Collaboration, others with this knowledge have sought to show how their discipline could be better structured. While Wellek and Warren reflected upon literary studies and distinguished the different tasks within it, Arne Næss is concerned with implementing something new in the relationships between humanity and nature. Scholars familiar with Lonergan's discovery reflected on the state of their particular disciplines – feminist studies, language studies, economics, and the natural and formal sciences – and saw the need for some order. In the field of comparative housing research, Michael Oxley proposed a division of labour for understanding housing in different countries. He also indicated the process whereby policies can be transferred from one country to another.

These are varying expressions of the search for a scientific approach to a discipline. To paraphrase McShane (2002: 57), the key point is that these various scholars, whether familiar with functional specialisation or not, are not inventing distinctions; rather, they are ordering distinctions invented by centuries of searching in different disciplines. These distinctions are now richly evident in journal articles that collect and organise data, that seek to interpret other scholars in their discipline, that grapple with the problem of putting in sequence not only the ideas of the past but also the events of the past, and that seek to implement a new future.

Notes

1 *Method in Theology* will be referred to from now on as *Method*.

2 See the series of Collected Works being published by the University of Toronto Press: http: //www.lonergan-lri.ca/store. As at June 2013, 19 of a projected total of 25 volumes have already been published. For an introduction to Lonergan's life, his leading ideas and his intellectual development, see Lambert and McShane (2010) and Mathews (2005).

3 As Marx ([1845] 2002) famously said: 'Philosophers have hitherto only interpreted the world in various ways; the point is to change it.'

4 *Insight: A Study of Human Understanding* will be referred to from now on as *Insight*.

5 Lonergan is not speaking of belief in a specifically religious sense. Rather, belief grounds co-operation between persons and thus what in the social sciences is referred to as socialisation or acculturation.

6 Lonergan understands 'conversion' not simply in a religious sense. In Chapter 7, I will refer to it as a displacement of horizons – from the horizon of common sense to the horizon of theory, and from the horizon of theory to the horizon of the subject as subject.

7 A biographical note is important here. Soon after this discovery of functional specialisation in 1965, Lonergan was diagnosed with lung cancer and nearly died while having a lung removed. So, *Method* was written at a difficult time in his life as he recovered from serious illness (Mathews 1998).

8 *Insight* was largely an exercise in intentionality analysis. The language, however, was that of faculty psychology.

9 Rene Wellek (1903–1995) was a Czech-American comparative literary critic. See http: //en.wikipedia.org/wiki/Rene_Wellek, viewed 12 March 2010

10 Austin Warren (1899–1986) was an American literary critic and author. See http: // en.wikipedia.org/wiki/Austin_Warren, viewed 12 March 2010

11 In Wikipedia, Charles Augustin Sainte-Beuve (1804–1869) is described as a 'literary critic and one of the major figures of French literary history' (http: //en.wikipedia.org/ wiki/Sainte-Beuve, viewed 12 March 2010), Kenneth Burke (1897–1993) as a 'major American literary theorist and philosopher' (http: //en.wikipedia.org/wiki/Kenneth_ Burke, viewed 12 March 2010) and R.P. Blackmur (1904–1965) as an 'American literary critic and poet' (http: //en.wikipedia.org/wiki/R_P_Blackmur, viewed 12 March 2010).

12 For a more detailed discussion of Oxley's article, see McNelis (2009b).

13 Including *Feminist Studies*, *Signs: Journal of Women in Culture and Society* and *Canadian Woman Studies*.

14 See also other articles in *Signs: Journal of Women in Culture and Society*, Special Issue: 'Feminisms at a Millennium', Summer 2000.

15 Gillis-Drage (2010: fn 13) refers to Magarey and Sheridan (2002) and to Thompson (2002) who 'provides a very good multiracial perspective on the same question'.

16 Gillis-Drage is quoting from Thompson (2002: 340).

17 'The Great Circle of Feminism' is a reference to the formation in 1910 of Bharat Stree Mahamandal (The Great Circle of Indian Women) (http: //en.wikipedia.org/wiki/ Bharat_Stree_Mahamandal by Sarala Devi Chaudhurani (1872–1945), an early Indian feminist, http: //en.wikipedia.org/wiki/Sarala_Devi_Chaudhurani, viewed 15 March 2010.

18 He has in mind Ray Jackendoff, Noam Chomsky, Steven Pinker, Jerry Fodor, Bernard Comrie, Jacques Derrida and John A. Hawkes (fn 9, p.6)

19 The ten major journals were *Americanist, Essays in Criticism, Cambridge Quarterly, Forum for Modern Language Studies, Literary and Linguistic Computing, Literature and Theology, Notes and Queries, Review of English Studies, The Year's Work in English Studies* and *Journal of Linguistics*.

20 See the discussion above.
21 A reference to the influential social theorist George Simmel who speaks about 'turning to the idea' or 'displacement towards system'

> to denote the tendency and even the necessity of every large social, cultural, or religious movement, to reflect on itself, to define its goals, to scrutinize the means it employs or might employ, to keep in mind its origins, its past achievements, its failures. Now this shift to the idea is performed differently in different cultural settings. While a historical tradition can retain its identity though it passes from one culture to another, still it can live and function in those several cultures only if it thinks of itself, only if it effects its shift to the idea, in harmony with the style, the mode of forming concepts, the mentality, the horizon proper to each culture. (Lonergan [1969] 1974: 159).

See also Simmel ([1918] 2010: Chapter 2, 19–61).
22 McShane refers to three articles by King (1993), De Vroey (2000) and Romer (2000). The IS–LM model is central to Keynesian economics. It refers to the relationship between two variables, interest rate and national income: IS (Investment/Savings) refers to national income and interest rate at which there is equilibrium in the market for goods and services; LM (Liquidity Preference/Money Supply) refers to national income and interest rates at which there is equilibrium in the money market.

> The IS specification is typically portrayed as downward sloping, which results from assuming that savings is a positive function of income and investment is a negative function of the interest rate (as the cost of capital). The LM schedule is typically portrayed as upward sloping since money demand depends positively on real income (as a measure of the scale of transactions) and negatively on the nominal interest rate (as an opportunity cost of cash balances). (King 1993: 69)

Thus, in Keynesian economics, the economy is in general equilibrium when these two graphs intersect (Black et al. 2009).
23 McShane refers to seven articles in *Economica* 67 (2000), particularly an article by Wen-Ya Chang and Ching-Chong Lai, 399–417.
24 McShane in this context refers to Functional Collaboration as 'Hodic Method'. 'Hodic' has various associations: the Indo-European *hod* refers to *way*; *hod* also has the meaning referred to in the line of the song *Finnegan's Wake*: 'And to rise in the world he carried a hod' (McShane 2002: 37); a *hod* is a builder's V-shaped open trough on a pole, used for carrying bricks and other building materials (Oxford Dictionary).

3 Towards a scientific approach to research

> Dynamically a science is the interplay of two factors: there are data revealed by experience, observation, experiment, measurement; and on the other hand, there is the constructive activity of mind. By themselves the data are objective, but they are also disparate, without significance, without correlation, without coherence. Of itself, the mind is coherence; spontaneously it constructs correlations and attributes significance; but it must have materials to construct and correlate; and if its work is not to be fanciful, its materials must be the data. Thus thought and experience are two complementary functions; thought constructs what experience reveals; and science is an exact equilibrium of the two.
>
> Bernard Lonergan in *For a New Political Economy* ([1942–1944] 1998: 5)

I am proposing that Functional Collaboration is a scientific approach through which we can make progress in housing. But what is a scientific approach to any subject? What distinguishes a scientific approach from any other approach? Can we have a scientific approach to social science? What is the status of scientific knowledge? These and many more fundamental questions about science have long been debated.[1]

At the beginning of his book, *The Beginnings of Western Science*, David Lindberg (1992) reviewed the many different meanings of the term 'science'. Different traditions describe it in different ways. For example, positivists highlight the empirical aspects; constructivists such as Schutz (1982) and Berger and Luckmann (1971) highlight the role of social agents in constructing the world; critical realists such as Danermark et al. (2002) and Sayer (2010) highlight the objective existence of reality in terms of observation and theorising; Kuhn (1970) and his successors highlight the historical and revolutionary aspects of science; others highlight its logical processes of induction and deduction (Mill [1882] 2009).

Many traditions reserve the term 'science' for the natural sciences rather than the social or human sciences. Some reserve the term to understanding the world rather than applying that understanding. Flyvbjerg (2001), on the other hand, dismisses the traditional understanding of science as 'epistemic science', arguing that social science cannot do what the natural sciences do. In his view, the key characteristic of the natural sciences is its capacity to predict. As social science cannot predict, it should abandon its goal of being an epistemic science. He proposes that social science should be a phronetic science. This is where social science is strongest. Phronetic social science deliberates about social actions, in particular, values and power. It operates in a context where variables are dependent upon the context. Flyvbjerg seeks to 'restore social science to its classical position as a practical, intellectual activity aimed at clarifying the problems, risks and possibilities' (2001: 4).

The common aim of all disciplines, regardless of their philosophic tradition, 'is to advance knowledge in their field, to provide new or better understanding of certain phenomena, to solve intellectual puzzles and/or to solve practical problems' (Blaikie 2003: 10–11). For the most part, disagreements about what is and what is not science revolve around differing descriptions of science. To my mind, these descriptions do not reach what is constitutive of science, what is most central, significant and relevant to an understanding of science. These descriptions focus on the characteristics of the methods and results of a discipline: whether the data collection is sound; whether its methods produce accurate, rigorous and precise results; whether the discipline can predict what will happen. Indeed, the importance of a scientific approach to housing research does lie in the results it produces. These results bear fruit not only in greater understanding of housing but also in the implementation of better housing practices. Science contributes not just to the advancement of knowledge but also to the advancement of better living. If science is to contribute to better living, this advancement of knowledge must extend to a better understanding of what is required to implement this living. However, getting results presupposes a method, a way of going about getting results. Method is 'a set of directives that serve to guide a process towards a result' (Lonergan [1957] 1992: 421). Further, method presupposes a question. It guides the process that takes us from a question to an answer (the result). Different methods answer different questions and, as we know, different methods produce different results.

So, the question arises, by what method do we find an answer to the question 'what is science?' or the question 'what is a scientific approach to housing?' As yet I'm not yet prepared to answer this question, for, as we will see in Chapter 5, asking and answering such questions are not as straight-forward as they appear. It is only in Chapter 9 that an *explanatory definition* of science will emerge. Here I want to lay some initial groundwork for later discussions by reflecting upon the questions we ask and the type of answer each question anticipates. After all, asking and answering questions is central to science and a scientific approach to housing. I outline a proposal for a scientific approach to housing research. I begin by outlining in very general terms a current standard model of housing research.

A current standard model of housing research: external and internal processes

Method is the process whereby we bring about some result. This result can be envisaged as some endpoint or as some interim point on the way towards that endpoint. So it is within housing research that an investigation consists of a series of steps, each of which reaches some interim result which contributes, in some way, to some endpoint result. Such a series of steps may include:

1   Defining the problem more clearly, which involves setting out a series of research questions whose answers are interim points to solving the problem.
2   Reading the research relevant to that problem in order to, among other things, (i) find out the experiences of others and their reflections on and understanding of the problem and (ii) work out ways in which the series of research questions could be answered. This too involves a series of steps, each of which produces some interim result such as searching out and determining the relevant research, prioritising what research will be read and to what extent, reading each book/article, assembling the results of each book/article, determining gaps in the research etc.
3   Undertaking some field research. This may include some combination of undertaking a survey, interviewing participants and observation. It also involves a series of steps, each of which produces some interim result such as drafting a set of questions, piloting these questions, engaging participants, distributing the survey or undertaking interviews or making observations, developing a framework of analysis, analysing the results and reaching some conclusions.
4   Writing up these conclusions. Again this involves a series of steps, each of which produces some interim result such as structuring the report into chapters, sections, paragraphs and sentences.

In describing this series of steps (and a series within each step), I am seeking here to outline a standard social science research model that is commonly, but by no means universally used in housing research. While quantitative and qualitative methods differ in their pre-suppositions, the type of data collected, whether hypotheses are proposed, the type of hypotheses proposed and the type of conclusions drawn, they both go through this series of steps. This standard model is common across different epistemological and ontological approaches whether positivist, interpretivist, critical realist etc. even though they strongly disagree about the starting point for housing research. It is a common across academic disciplines. It is common across research undertaken for different purposes whether theoretical, policy analysis, strategic planning etc.[2]

As described here, this standard model involves a series of external activities. Whether it is an adequate description, whether it adequately covers different approaches, disciplines and purposes may be a matter of some debate. Here, however, I want to draw attention to something beyond these external activities: for each external activity there is a parallel internal activity, whether experiencing, imagining, thinking, considering, supposing, formulating, defining, judging, questioning, criticising, evaluating, deciding. So, the doodling, drawing, sketching, writing etc. that is involved in defining a problem, or the reading and note-taking involved in reading others' research, or the analysis of field research, has a parallel set of internal activities that guide the external activities within each step, such that each step is achieved and contributes to reaching some endpoint of the overall research project.

Further, as the housing researcher proceeds through the external activities, the internal activities may change the researcher such that they understand and define the problem in a new way. As they do so, they change the external activities – they read other research; they do other field research, they interview different people, they interrogate different datasets; they reach different conclusions. These internal activities are not just elements of the rational processes of logic. They are also elements in the processes of understanding, of learning, of discovery, of creativity and of innovation, processes that shift from one frame of reference to another. At times these processes achieve results slowly and incrementally. At other times they can achieve them rapidly in leaps and bounds. Transformations in understanding can occur at any time as the researcher proceeds through the series of external activities. These transformations require the researcher not just to change the external activities but to begin again on a different basis. So it is through some combination of these internal and external activities that we make progress in understanding a problem (and its solution).

At times these internal activities, however, are not observable. Yet as housing researchers we are more or less conscious of them. Indeed, with practice we can become more conscious of them. Nevertheless, many housing researchers are so focused on a housing problem and coming up with an immediate practical solution to this problem that they are largely unaware of the internal processes of research, do not take account of them or set them aside. For these researchers the significance of the data and the solution(s) they propose are largely unproblematic. Their results are continuous with the work of practitioners seeking solutions to everyday problems. This common-sense approach to housing research operates within the taken-for-granted presuppositions of a culture and seeks to solve the immediate problems of practitioners.

More fundamentally, however, housing researchers operating within this common-sense approach to housing presuppose the adequacy of their internal processes. Yet, the same internal processes that lead a researcher to ask about the significance of data and possible solutions to everyday problems can also turn inward and raise questions about these internal processes: what precisely is being investigated? are the answers to my questions adequate? by what criteria do I judge that I have reached an adequate answer? to what extent is my perspective limited by involvement and interests? etc.

It is through a phenomenology of research that we can come to some understanding of our internal activities, some self-understanding of what we are doing when doing research. To do so may require us to go beyond the limitations of traditional phenomenology, viz. that it has not sufficiently explored rationality. 'Phenomenology is concerned with what is evident. The thematic treatment of what is evident is considered secondary, merely the phenomenologist's report. But there has not been done a phenomenology of rational consciousness, of the process of asking, Is it so? ...' (Lonergan [1957] 2001a: 275).

Kemeny urges housing researchers to take conscious control of the research agenda and 'acknowledge, and take a stand on, the inevitability of their involvement in the social construction of housing issues' (1992: 32). Here, I am suggesting that we also take conscious control of the processes of research. If we want to improve our chances of developing good research, then it will help to understand what goes on when we are researching. Once we understand what goes on, then we can take some deliberate control over how we proceed and over the varied but related activities that constitute research (to paraphrase Shute and Zanardi 2006: 31).

Most housing research is locked into a taken-for-granted everyday common-sense approach to housing research. To shift out this approach, a housing researcher must advert to their internal processes and ask a key question: what am I doing? This is not an easy question to answer.

In *Housing and Social Theory* and elsewhere, Kemeny has called for a shift to theory in housing studies. What such a shift might mean depends upon an answer to the question: what is theory? It is this question that Peter King (2009: 51) asks at the conclusion of his article on 'Using Theory or Making Theory': 'Perhaps this paper should be taken as a challenge to established approaches to housing and social theory: as a questioning of what we mean by theory; what its possibilities and limits are; and, in consequence, what we take housing to be.'

In Chapter 7 this question about theory becomes: what am I doing when I am theorising? Answering such a question, however, is not enough. A shift to theory in housing studies depends upon a decision of the housing researcher to act in accordance with that answer. It is this decision that shifts a researcher into the world of theory. I would understand this shift as more of a 'displacement' into a new world rather than a shift that is continuous with the everyday taken-for-granted world of common-sense living.

A decision to shift into the world of theory, however, presupposes an answer to another question, one which evaluates whether it is worthwhile doing so. To advert to this presupposition raises the question: what am I doing when I am evaluating?

These 'what am I doing?' questions presage a shift into another world: the world of the subject as subject. I will discuss 'the subject as subject' more fully in Chapter 7 and propose not just a single displacement into the world of theory but rather a double displacement: a displacement into the world of theory and a more radical displacement into the world of the subject as subject.

Here, however, I want to highlight the gap between the current culture of housing research and what I will propose in Part II. This gap presents a difficulty in communicating a new understanding (and doing) of science.

For complete self-development is a long and difficult process. During that process one has to live and make decisions in the light of one's undeveloped intelligence and under the guidance of one's incomplete willingness. And the less developed one is, the less one appreciates the need of development and the less one is willing to take time out for one's intellectual and moral education.

(Lonergan ([1957] 1992: 650)

For a housing researcher the fundamental challenge is to reach some self-understanding of what they are doing when doing research. It involves a long, difficult and personal journey of understanding oneself, of understanding their practices and their methods, of understanding what they are doing when they are investigating, theorising, evaluating or deciding to implement something new.

Any new scientific discovery whether in quantum physics, chemistry, biology or the social sciences is the result of a personal journey of discovery. By striking out on new paths individuals have brought about revolutions in science and innovation in everyday living. Here the personal journey of discovery is a journey within. It is just as difficult as any other scientific journey, if not more. It is just as difficult to communicate.

> Just as the non-physicist can identify and distinguish the colours of the rainbow, so the philosopher may identify and distinguish types of understanding. But such identification by the philosopher is no more the generalized empirical method that is generative of scientific metaphysics than descriptive identification of types of flowers is scientific botany…
>
> This notion of a generalized empirical method, a self-attentive pursuit which has the dimensions of a science, may be unpalatable to some, incredible to others. But on its operative admission stands or falls the project of ongoing collaboration. One may find…the projected self-attention obscure. If so, one must begin from precisely that discovery: 'He can contrast that experience of not understanding with other experiences in which he felt he understood. Then he can turn his efforts to understanding his experiences of understanding and not understanding. Finally, when proficient at introspective understanding, he can move to the higher level and attempt to understand his successful and unsuccessful efforts at introspective understanding'. The entire project is no mean task…It is concretely emphasized…by indications of time and lifetime. I have spent many months struggling self-attentively towards an understanding [of experiences of understanding]. I have spent several years introspecting my insights in physics, chemistry and biology in order to bring me to some appreciation of the autonomy or non-reducibility of botany.
>
> Obviously, if there are scientific insights within this area of self-attention, they are no more communicable in brief than the insights of Quantum Theory can be communicated in a short course in physics.
>
> (McShane 1976: 11–12).

Research as asking and answering questions

One starting point is to consider our questions. All housing research begins with a researcher asking some sort of question. While this question is about something, the question itself has its origins in the researcher. The question emerges within a context of other larger questions and an answer to this question will make some contribution to these larger questions. On the other hand, the question guides a particular research project and within it lays a series of secondary questions whose answers will provide some contribution to answering this question.

Method presupposes a question to be answered. And different methods are used to answer different questions. Unfortunately, as William Mathews (1987: 249) comments: 'Philosophers of science, with their concern with verification and falsification, or the social framework of scientific inquiry, have not paid nearly enough attention to the anticipatory structures of scientific questions.'

Books on research methods and the philosophy of social science devote little attention to 'questions' (apart from questions within questionnaires) and rarely include a reference to 'questions' in their indexes (see, for example, Flick et al. (2004), Danermark et al. (2002), Winch (2003) and Sayer (2010)). De Vaus (2002) has some initial discussion of 'research questions' but not in relation to the anticipatory structure of different types of questions. As discussed below, Blaikie (2003: 13ff; 2007: 6f) is one exception.

In our everyday living, we ask all sorts of different questions which emerge from within us as we puzzle about something. We are puzzlers, and questions as thought and formulated emerge as expressions of this puzzling, as expressions of our desire to understand and to solve problems. They reveal something about our dynamic orientation towards understanding and responding. They emerge in response to something that we are interested in or care about or are puzzled by.

As in everyday living, housing researchers ask all sorts of questions about different topics. For instance, Jacobs et al. (2010b) in an AHURI research project on the future of public housing ask:

> What problems have arisen for the SHAs as a result of trying to manage housing stock in a period of tight budgetary constraints?
>
> How have the major drivers shaping public housing provision (social, economic and political) affected its future role?
>
> How might or should the SHAs prepare for the future policy environment to ensure that funds spent on public housing achieve sustainable policy outcomes?
>
> To what extent is public housing viable in the current policy environment?
>
> What alternative models of provision might be used in the future to enhance the broader role of public housing?

Or again, McNelis et al. (2008) in another AHURI research project on older persons in public housing ask:

> What are the characteristics and housing circumstances of older public housing tenants?
>
> What is the likely future demand for public housing from older persons over the next ten years?
>
> What are the housing policy and management issues associated with older tenants?

What is the role and responsibilities of SHAs in facilitating the access of older people to support services, in particular, to aged care?

What examples of good practice and policy initiatives are there among social housing providers in Australia and overseas?

Like these two examples, a research project will consist of different sorts of questions about a particular topic. What is happening? Why has it happened? Who are the major players and what are their characteristics? Who is doing what? What are the current issues or problems? How can we make things better? How is power being exercised? How has something changed? What is its history? What is driving change? What are its prospects for the future? Who is responsible for what has happened and for what might happen in the future? etc.

Asking questions in the context of everyday living (where immediate practical responses are called for) is very different from asking questions in a scientific context. In the first place, in the context of everyday living we are apt to ask the same question in different ways. In this context, this is acceptable. However, in a scientific context, we need to be more precise about the questions we ask. In the second place, in the context of everyday living we are apt to focus on one particular question or even a couple of questions, either to the exclusion of other questions or without recognising how one relates to another.

By focusing on and enquiring into the anticipatory structure of questions we can be much more precise about (i) the intent of our question and (ii) the relationship between one question and another question, and we can (iii) work towards a complete set of questions.

Norman Blaikie in *Analyzing Quantitative Data: From Description to Explanation* (2003) proposes that:

> Research questions are of three main types: 'what' questions, 'why' questions and 'how' questions:
> 'What' questions seek descriptive answers.
> 'Why' questions seek understanding or explanation.
> 'How' questions seek appropriate interventions to bring about change.
> All research questions can and perhaps should be stated as one of these three types. To do so helps to make the intentions of the research clear. It is possible to formulate questions using different words, such as, 'who', 'when', 'where', 'which', 'how many' or 'how much'. While questions that begin with such words may appear to have different intentions, they are all versions of a 'what' question: 'What individuals...', 'At what time...', 'At what place...', 'In what situations...', 'In what proportion...' and 'To what extent...'. (p.13)

For Blaikie, social research can pursue one of a set of eight objectives – explore, describe, understand, explain, predict, change, evaluate and assess social impacts – with the first five being characteristic of basic research and the last three of applied research. These follow a sequence, with the 'core of all social research [being] the sequence that begins with the *description* of characteristics and patterns in social phenomena and is followed by an *explanation* of why they occur' (pp.11–12, see also Blaikie 2007).

If, however we consider our questions more closely, we will find that a range of different types (according to the answer they anticipate) emerge from within us as we puzzle about something.

- what-is-it-questions anticipate that there is something to be understood, something to be made sense of. They anticipate an insight into something within our experience such that we will reach some understanding of the 'what' under consideration;
- purpose-questions anticipate that this 'what' serves some purpose or role;
- is-it-questions seek a judgement as to whether or not this particular event is the occurrence of a 'what';
- change-questions anticipate an understanding of what is going forward in the development of the 'what' under consideration;
- how-often-questions anticipate a frequency, the occurrence of the 'what' so many times within a group or class or population of events;
- is-it-worthwhile-questions seek an evaluation as to whether it is worth the effort creating this 'what';
- what-do-I-do-questions anticipate a course of action within which the 'what' is not simply understood but will be created.

While Blaikie specifies only three types of research questions, this indicates a broader range. Many researchers highlight why-questions as seeking an explanation and Blaikie notes that 'some questions that begin with 'what' are actually 'why' questions. For example, 'What makes people behave this way?' seeks an explanation rather than a description. It needs to be reworded as: 'Why do people behave this way?'' (2003: 13). So, it is notable that none of the questions in the above set asks a why-question. This is because why-questions intend different answers and we need to distinguish between different types of why-questions. Why is a house, a house? (i) because it consists of certain materials – bricks, timber, concrete etc., (ii) because these materials are ordered in a certain way, (iii) because it was built by a group of people and (iv) because it meets some purpose (McShane 2003; see also Aristotle [*c*.322BC] 2007b: Book II, Section 3). These different answers to a why-question point to different types of causes: material cause, formal cause, efficient cause and final cause.[3] As such, why-questions reflect different types of questions.

The questions above are not a definitive list. It still needs some refinement, but it does begin to sort questions into their basic types and so, we can be more precise about our questions and the answers we are anticipating. Moreover, we can note that some questions presuppose some answer to others, i.e. that there is some order and some relationship between questions. For instance, all these questions presuppose an answer to a what-is-it-question. So, before we can answer a what-do-we-do-question to create housing, we need to know what housing is, we need to know what purpose or role it plays and have decided that it is worthwhile promoting or creating; before we answer an is-it-worthwhile-question, we need to know what is housing that we are deciding about; before we answer a how-often-question or an is-it-question, we need to know whether it is housing that is occurring and distinguish its occurrence from the occurrence of something else. A what-is-it-question, then, provides a context for answering further questions.

A proposal for a scientific approach to housing research

In his introduction to *Housing and Social Theory*, Kemeny (1992: xviii) referring to 'a sociology of residence' notes that

> [t]heorising is not merely the mechanical application of ideas from one field to another. It is essentially the use of what C Wright Mills called 'the sociological imagination': calling into question aspects of social structure that tend to be taken for granted and using fantasy and lateral thinking in the craft of sociology to solve problems, often reformulated in novel ways.

And as Herbert Marcuse notes: 'Without fantasy, all philosophic knowledge remains in the grip of the present or the past and severed from the future, which is the only link between philosophy and the real history of mankind' (1968: 155).

Using 'fantasy and lateral thinking' in relation to a scientific approach to housing research, I propose to take the methods of housing researchers and reformulate them in a novel way. So, Functional Collaboration will be discontinuous with past understandings of housing research. Thus, in some sense, Functional Collaboration could be described as a fantasy, a vision of the future, an anticipation of something new. Here it may be worthwhile keeping in mind some of the great discoveries of the past. For example, Mendeleev discovered the periodic table in chemistry. Prior to this event, chemical processes already operated and chemists had developed some understanding of these processes. Mendeleev, however, through 'fantasy and lateral thinking' leapt to a new understanding and formulation of the whole of chemistry which incorporated all the elements, their structure and their relationship to one another. Once understood, the periodic table became the foundation for modern chemistry and opened up a vast range of possibilities to human ingenuity.

Functional Collaboration is not a fantasy in the idealist sense of dreaming of some future utopia. Rather, it is an attempt to formulate a discovery of something that already operates (albeit confusedly) – how human history brings about progress – the elements, their structure and their relationship to one another. An understanding of this discovery anticipates opening up a vast range of possibilities in any area of human endeavour.

So, what follows is an invitation to 'fantasy and lateral thinking' that operates on two levels. On one level, it requires the housing researcher to have some understanding of the variety of methods used in housing studies; it requires the housing researcher to ask: what am I doing when I am using this method? how does this method relate to other methods? and, how can I integrate these various methods into a single whole? On a second level, it requires a housing researcher to pay attention to the type of question underlying each method and so begin (i) to differentiate different types of questions, (ii) to grasp what each question anticipates, (iii) to grasp what they are doing in the process of answering each type, i.e. the method by which an answer is sought to each question, (iv) to relate each type of question to the others and (v) to affirm that it is by seeking new answers to all these questions that we can make some progress in understanding some housing issue and in understanding the method by which we reached this understanding.

In housing research we can ask a range of different types of questions and here I want to propose that a scientific approach to housing would seek to incorporate all the questions that could be asked of housing. It would be a complete set of questions; the questions would have a certain order; the questions would be inter-related with each question understood in the context of all the other questions and thus in relation to one another. Further I want to propose that each type of question is a stage in the process from understanding the current situation to implementing something new; that each type of question anticipates a particular type of answer – it has a defined purpose; that each type of question is answered through a particular method; and that the question and purpose of each stage provides the grounds for a division of labour into functional specialties and for a new form of collaboration between housing researchers. Table 3.1 outlines the eight functional specialties, the corresponding type of question that each functional specialty seeks to answer through a particular method and a description of the corresponding purpose of each functional specialty.[4]

Table 3.1 Functional specialty, type of question answered and a description of its purpose

Functional Specialty	Type of questioned answered	Description of purpose
Research functional specialties: understanding the past		
Research	Empirical question	Gathering and making available the relevant data
Interpretation	Theoretical question	Understanding what is being investigated
History	Historical question	Understanding what is going forward
Dialectic	Evaluative/critical question	Coming up with the best story and the best basic directions

Table 3.1 continued

Functional Specialty	Type of questioned answered	Description of purpose
Implementation functional specialties: looking to the future		
Foundations	Transformative/ visionary question	Expressing the best fundamental directions – working towards a better basis upon which researchers can operate
Policies	Policy question	Reaching for new guidelines and directions for change
Systematics	Strategic question	Integrating possible directions for change with other theories/other aspects of society
Communications	Practical question	Working out specifically how strategies can be applied within the local context of roles and institutions.

Table 3.1 describes the different parts of a scientific approach to housing. Rather than a descriptive definition of science, however, Part II works towards a more precise understanding of each functional specialty while Part III shows the relationship between the functional specialties such that functional specialties constitute a theory or explanatory definition of science.

Notes

1   See, for example, Husserl (1970), Popper (2002), Rorty (1989), Feyerabend (1993), Kuhn (1970), Lakatos (1970, 1978), Feyerabend (1993), Winch (2003), Giddens (1979, 1993), Bhaskar (2008), Blaikie (2007) and Sayer (2010).
2   Crotty (1998: 2), for example, argues that 'four questions are basic elements of any research process': What *methods* do we propose to use? What *methodology* governs our choice and use of methods? What *theoretical perspective* lies behind the methodology in question? What *epistemology* informs this theoretical perspective?. Answers to these questions will determine the specifics of this standard social research model that I have outlined here.
3   In his *Posterior Analytics* ([*c.*322BC] 2007a: Book II, Section 1), Aristotle discusses different types of question. He also discusses how an understanding of the *whatness* of something is the same as understanding the *why* of it (see his discussion of understanding the eclipse of the moon in Book II, Section 2).
4   These terms (with the initial letter in upper case) have a technical meaning. As functional specialties, 'Research' and 'Policies' should not be confused with how these terms are generally used. Their more technical meaning is outlined in Part II.

Part II

The functional specialties

Dialectic   Foundations

History   Policies

Interpretation   Systematics

Research   Communications

Chapter 3 proposed that a scientific approach to housing will be a complete set of eight inter-related questions. Part II seeks to develop a more precise understanding of each of the eight questions and the method by which each is answered. A refined understanding of these questions is a long and gradual process, yet essential to understanding Functional Collaboration as a scientific approach that will provide practical advice that is a condition for bringing about progress in housing.

The initial problem of pre-suppositions

In presenting Functional Collaboration, I must begin with some understanding of my audience. So, rightly or wrongly, I presuppose an audience of housing researchers operating within the everyday taken-for-granted world of common-sense living. So, immediately, I am confronted with a gap in world-views, for an understanding of Functional Collaboration presupposes the 'displacement' of housing researchers into not only the world of theory but also into the world of the subject as subject. A housing researcher operating in the everyday world of common-sense living has little idea of the world of theory and the world of the subject as subject.

It would be easier to bridge this gap if I could point to instances where Functional Collaboration had been explicitly adopted and implemented. As noted in Chapter 2, however, this is not possible.

If this were not enough, presenting a precise understanding of each functional specialty presents a further problem: the presentation itself already presupposes an understanding of Functional Collaboration as a whole. Yet, to communicate an understanding of Functional Collaboration as a whole presupposes an understanding of all the parts, their relationships to one another and their relationship to the Functional Collaboration as a whole.[1]

The problem I face is illustrated in the following quote from Patrick Byrne (2002: 21) regarding the functional specialty Research; the same problem applies to any of the other functional specialties as well.

> The term 'research' is used widely and in various ways in our contemporary world, especially in university contexts. In this undifferentiated sense, research covers a vast range of very different activities: conducting laboratory experiments in the natural sciences, searching for information in libraries or on the internet, performing statistical analyses of data in the social sciences, discovering new theoretical principles, devising new proofs of problematic theorems, constructing theoretical models, deconstructing literary texts, and constructing critical historical accounts. On the other hand, Bernard Lonergan restricted his use of the term Research to a very technical sense in *Method in Theology*. There Research refers strictly to the activities involved in one of his eight functional specialties.

My problem? The term 'research' along with the term 'policy' is most often used in a general sense within housing studies, in what Byrne refers to as an 'undifferentiated sense'. As it is generally used within housing studies, 'research' tends to cover in an imprecise way the first four of the functional specialties. Within Functional Collaboration, however, the term 'Research' is used in a precise technical sense as one of eight functional specialties. It is understood and undertaken 'in terms of its functional relations to the other seven functional specialties' (Byrne 2002: 24).

Prior to any attempt at presenting Functional Collaboration, I have already come to my own understanding of Functional Collaboration as a whole and as constituted by each of its parts, the eight functional specialties. This understanding of the whole and the parts is presupposed in my presentation.

Two strategies seek to address this problem of pre-suppositions. The *first strategy* is to structure the discussion of each functional specialty as a movement from the everyday taken-for-granted world of common-sense living to a more precise understanding of that functional specialty. So, the discussion follows a six-stage pattern (with some variations). First, it begins with an appeal to the experience of a researcher undertaking research as an initial view of the functional specialty. This phenomenology of research highlights some aspect of research. It asks the reader to reflect upon their experience of doing research, distinguishing one element of this experience from another. This phenomenology of research orients the discussion that follows and an understanding of each functional specialty hinges on the capacity of the reader to distinguish elements within their own experience of doing research. Second, this phenomenology of research is followed by a review of genres of housing research relevant to the functional specialty. Third, a key question is asked of the relevant genres within a functional specialty: what are the authors in these genres doing? So, this stage presents a brief summary of their significant elements. Fourth, an attempt is made to develop a more precise understanding of the functional specialty. Fifth, for the functional specialties Interpretation, History and Dialectic there follows an illustration. Sixth, each chapter concludes with a formulation of the question underlying the functional specialty as well as some comments on its implications for housing research.

Functional Collaboration presupposes a particular understanding of the world of theory, of the world of the subject as subject, and of the exercise of power whereby the interests of particular groups dominate a society. These are key issues for housing research and a scientific approach to housing must address them. The *second strategy* discusses these issues as they arise within the functional specialties. Interpretation distinguishes theoretical understanding from the understandings of everyday common-sense living. Dialectic discusses the conflicts that arise as housing researchers operate in the everyday world of common-sense living and deliberately or implicitly operate within the horizon of one or other interest group. It also discusses the conflicts that arise when housing researchers have not decisively shifted into the world of theory. Foundations discusses the decision facing housing researchers as to which is the best way forward in conducting their research.

A clarification

Within housing research and, more broadly, within the history of science and method, there are many highly contestable issues, each with a very extensive literature, each subject to wide and intense debate, each with a long and complex history. In recent decades, among many others, Karl Popper (2002), Richard Rorty (1989), Thomas Kuhn (1970), Imre Lakatos (1970, 1978), Paul Feyerabend (1993), Anthony Giddens (1979, 1993), Norman Blaikie (2007), Peter Winch (2003), Roy Bhaskar (2008) etc. and their protégés have all put forward views on science and method raising a raft of issues and disputes.

All these many different and conflicting views deserve to be taken seriously. Before we can critique them, texts have to assembled, interpreted and located within the history of science and method. Chapter 4 on Research, Chapter 5 on Interpretation and Chapter 6 on History, however, point to the difficulties of doing this adequately. Moreover, as proposed in Chapter 7 on Dialectic, it is not an easy task to address the differences and conflicts between these various authors and to reach a viewpoint of one's own. It is simply not good enough to juxtapose one framework (that of the original author) with another (my own). Dialectic demands much more.

So here I do not seek to present an understanding, a history, a critique and evaluation of other views on housing research, science or method. I do not attempt to show how Functional Collaboration integrates these various views. A fuller view of Functional Collaboration would deal more thoroughly with different philosophical approaches (cognitional theory, epistemology and metaphysics), different ethical stances (values, means/end, is/ought) and different understandings of sociality and intersubjectivity, history and progress, and science and scientific method. Further, I do not discuss whether the implementation of Functional Collaboration is worthwhile nor discuss how Functional Collaboration could be implemented in housing research.

My goal here is simply to indicate through an exploration of the subject as subject and, in each chapter, through a phenomenology of research that there are different questions, that different questions anticipate different types of answers and appeal to different types of evidence, that seeking answers to these different questions requires different methods, that these questions and methods are related and are parts of a larger whole, a view of science which can contribute to progress in housing. To do this adequately is a huge yet essential task.

As Part II and Part III unfold, it will become clear that I am presenting an understanding of science. In doing so, I am operating within only one functional specialty, Interpretation. I am 'practising what I preach'. I am implementing Functional Collaboration by operating within one functional specialty and answering one question. So, I abrogate any totalitarian ambitions and seek only to make a contribution to the whole area of disputes and conflicts about science and method by providing a framework, a heuristic for addressing the broader range of issues that bedevil science. Without a theory of science, we do not have a framework or heuristic to understand adequately the history of housing research and how it has changed or is changing. Without an answer to this key question, we cannot proceed to deal with differing and conflicting views emerging in history nor do we have a framework to evaluate and critique housing research. Without this understanding we cannot make a judgement about alternative views, we cannot make a decision as to whether they are worthwhile and we cannot implement our decision on them. I seek only to point to the possibilities and to the work that still remains to be done. For this reason I eschew any critique. Indeed, any 'critical' comments should be understood as a contrast which will highlight how an understanding of Functional Collaboration as science differs from those which currently prevail.

A note on the literature

Nevertheless, the range of literature that could be considered is very large indeed, for I touch upon many areas of social science – the nature of science, method in science, research methods, theory, history, critique, evaluation, comparative studies, epistemology, policy analysis, policy development, strategic planning, implementation, economics, sociology and cultural studies. I have, thus, been compelled to contain the range of literature considered. As I am primarily concerned with a scientific approach to housing, within the wider discussion of methods and each functional specialty, I have largely focused on those methods as they have expounded within the corpus of housing research. Consequently, I have drawn selectively on the broader literature within social science to illustrate a particular point and to clarify, by contrast, the position I propose compared with other positions.

As noted above, the discussion of each functional specialty includes a review of relevant genres of housing research. Here I draw upon a range of Australian and international housing research including books, articles, reports, policy documents, manuals and other written material and have divided this research into particular genres according to their relevance to the discussion of a functional specialty: primary research (including the traditional quantitative and qualitative works); works which define housing (or some aspect of housing); works which propose analytical frameworks; histories of housing; critiques of housing; evaluation studies; comparisons between countries; works advocating particular philosophical positions; works on research methods; works which aspire to or envisage a new future for housing; works developing policy; strategic planning works; and works such as practice manuals. This division is guided by the functional specialties, and particular genres are allocated insofar as their basic orientation has some similarity to the initial view of a functional specialty outlined in the phenomenology of research. The allocation of works is not always clear-cut as authors shift rapidly from what I would locate in one specialty to that of another, not just within sections but from sentence to sentence. So, continually these authors shift back and forward from what is occurring, to its significance, to what is changing, to whether that is good or bad, to how to improve what is occurring. They are operating within the everyday world of common-sense living and have yet to adopt a scientific approach to housing. On the other hand, the different genres of research highlight the way in which different authors are doing different things, albeit in some muddled way. It is to the core of their work that I wish to point, as indicative of a functional specialty. So, the review of relevant genres seeks to highlight some aspects of what they are doing as significant and relevant, and to overlook other aspects as insignificant and irrelevant.

I have not attempted to cover all the relevant housing literature. Rather I have sought to indicate the types of literature that might be included in one or other genre. Further, as my knowledge of housing has developed within a particular context, the reader will find that, at times, the selected literature is skewed towards research on social housing in one state within one country – Victoria, Australia. This is particularly so in relation to some of the implementation functional specialties where examples of housing research within these genres are more difficult to find; they tend to be found within the 'grey literature' and are specific to an organisation. I trust that the reader who is not familiar with housing research or social housing research within Australia will be able to find parallels within their own country or within their own social science discipline.

A limitation

The eight functional specialties are divided into two phases. The first four (the research functional specialties) seek to understand the past. Research is discussed in Chapter 4; Interpretation in Chapter 5; History in Chapter 6; Dialectic in Chapter 7. The second four (the implementation functional specialties) look to the future and are discussed in Chapter 8.

All the implementation functional specialties are discussed in a single chapter whereas the research functional specialties have a chapter devoted to each. This may appear somewhat lopsided, particularly when housing research has a strong policy focus. It does, however, reflect the huge initial challenge facing housing research which currently operates largely within the everyday taken-for-granted world of common-sense living. This initial challenge is, as Kemeny urges, a transformative shift to the horizon of theory. Where attempts have been made to shift towards a more theoretical understanding, these, in my view, have not made a decisive break with the world of everyday common-sense living. As I argue in Chapter 7, such a break is not possible without the transformative shift to the world of the subject as subject. By placing theory within the larger framework of Functional Collaboration, I propose a notion of theory that differs substantially from that of Kemeny (and many others). At the same time, it critically distinguishes theory from Research, from History and from Dialectic while relating each of these to theory. Without these distinctions, the nature of theoretical understanding will continue to be confused and housing researchers will continue to retreat to the everyday world of common-sense living.

There is, however, a further personal reason for the lopsidedness in the discussion of the implementation functional specialties. It reflects my own progress in understanding Functional Collaboration and the difficulty of envisaging the shape of the implementation functional specialties as the research functional specialties shift into the horizon of theory. As someone who, for the most part of their career, has undertaken housing research from within the everyday world of common-sense living, to use 'fantasy and lateral thinking' to grapple with the possibilities of implementation within the horizon of theory was a bridge too far.

Note

1   This problem is referred to as the hermeneutic circle (Heidegger 1973; Gadamer 1975). Lonergan sets out both the problem and 'the answer' as follows:

It is understanding that surmounts the hermeneutic circle. The meaning of a text is an intentional entity. It is a unity that is unfolded through parts, sections, chapters, paragraphs, sentences, words. We can grasp the unity, the whole, only through the parts. At the same time the parts are determined in their meaning by the whole which each part partially reveals. Such is the hermeneutic circle. Logically it is a circle. But coming to understand is not a logical deduction. It is a self-correcting process of learning that spirals into the meaning of the whole by using each new part to fill out and qualify and correct the understanding reached in reading the earlier parts. ([1972] 1990: 159)

4 Research and the empirical question

Dialectic   **Foundations**

History   **Policies**

Interpretation   **Systematics**

Research **Communications**

[The experiential] is the given as given. It is the field of materials about which one inquires, in which one finds the fulfilment of conditions for the unconditioned, to which cognitional process repeatedly returns to generate the series of inquiries and reflections that yield the contextual manifold of judgments…the given is unquestionable and indubitable. What is constituted by answering questions, can be upset by other questions. But the given is constituted apart from questioning; it remains the same no matter what the result of questioning may be…the given is indubitable…what can be doubted is the answer to a question…But the given is not the answer to any question; it is prior to questioning and independent of any answers…the given is residual and, of itself, diffuse. It is possible to select elements in the given and to indicate them clearly and precisely. But the selection and indication are the work of insight and formulation…the field of the given is equally valid in all its parts but differently significant in different parts…some parts are significant for some departments of knowledge and other parts for other departments…It includes not only the veridical deliverances of outer senses but also images, dreams, illusions, hallucinations, personal equations, subjective bias, and so forth.

Bernard Lonergan in *Insight: A Study of Human Understanding*
([1957] 1992: 405–406)

Alfred Schutz notes that 'the *starting point* of social science is to be found in ordinary social life' (1972: 141). This is the taken-for-granted everyday world in which we live and carry on our day-to-day affairs. One of the elements in this taken-for-granted everyday world is housing. In this world, we take for granted the ways of understanding and doing:

It means to accept until further notice our knowledge of certain states of affairs as unquestionably plausible…Common sense thinking simply takes for granted, until counterevidence appears, not only the world of physical objects but also the sociocultural world into which we are born and in which we grow up. This world of everyday life is indeed the unquestioned but always questionable matrix within which all our inquiries start and end.

(Schutz 1982: 326–327)

This world is one 'which existed long before our birth, experienced and interpreted by Others, our predecessors, as an organised world. Now it is given to our experience and interpretation' (Schutz 1982: 208).

For Schutz:

[t]he structure of the social world is meaningful, not only for those living in that world, but for its scientific interpreters as well. Living in the world, we live with others and for others, orienting our lives to them...[The social scientist's] data...are the already constituted meanings of active participants in the social world. It is to these already meaningful data that his scientific concepts must ultimately refer: to the meaningful acts of individual men and women, to their everyday experience of one another, to their understanding of one another's meanings, and to their initiation of new meaningful behaviour of their own.

(Schutz 1972: 9–10)

This taken-for-granted everyday world of meaningful events is the starting point for the functional specialty Research as it seeks to answer the empirical question.

Research: an initial view

In his book *Social Research Methods: Qualitative and Quantitative Approaches*, Lawrence Neuman describes social research as 'a way of going about finding answers to questions' (2006: 2). All research begins with a question. It may be formulated or it may be just a sense of being puzzled. Whether formulated or not, my question is about something, for example, housing. This something (housing) precedes my question and provokes the puzzlement that issues in my question. Further, my question issues in an inquiry or investigation which, initially, turns to what I currently understand about this something (housing) but looks beyond this. While seeking something more, my inquiry turns to occurrences of this something (housing), to the context within which it occurs, to events that occur in the same time and place and to events that precede or follow it.

In doing this, I turn to what is given, to what precedes my question. I turn to the everyday taken-for-granted world of common-sense living, the world constituted by individuals who continually seek ways in which to express the meaning of themselves, of their relationship with one another and of their relationship with the world. Their meanings find expression in the materiality of the world. These expressions encompass both external events (states or actions) and internal events (states, actions, experiences etc.). They include attitudes, beliefs, opinions, feelings, emotions, values, habits, expectations, motivations, skills, capacities, personal and social characteristics, material characteristics (of buildings, of habitats, of environments), language, clothes, decorations, art, music, sounds, dance, performance, video, film, symbols, signs, customs etc. This diverse manifold of expressions as meaningful events is the starting point for social science (and for housing studies).

Schutz, through the work of Peter Berger and Thomas Luckmann (Berger & Luckmann 1971), is very much associated with qualitative research (and the social constructionist tradition of sociology) (Ritzer 2003: 373). Meaningful events in this taken-for-granted everyday world cannot only be described by participants and observed and recorded by researchers but they can also be counted and associated with other meaningful events. So, signing a lease is a meaningful event but it is one in which the number of times it occurs within a given time-period and geographical area can be counted; it is also one which can be associated with other events (such as a level or range of incomes) within the same time-period or geographical area.

In spontaneously asking and answering questions, I can distinguish two movements: a first movement turns to the field of materials, to what is given (the data) which provoked the question and about which we inquire; the second movement is the discovery of an answer to the question. By asking my question (and possibly a series of subsequent questions) about what is given, I move towards some discovery of the meaning or significance of what is given.

On an initial view, then, Research is concerned with what is given, with the field of materials about which we ask a question and whose significance is yet to be discovered.

Primary housing research

A review reveals three genres of housing research that have as their central orientation the first movement in asking a question, the turn to the occurrence of events. These genres make available different types of data, data that is turned to, attended to and gathered in response to a question. These genres are commonly referred to as primary or field research. They include some component of gathering the data that is relevant to an understanding of housing. This activity is the most prevalent within housing research. The difficulty, as we will see, is that researchers do not advert to the distinctions between gathering data, interpreting data, noting changes in data and critiquing or evaluating data.

Initially, when speaking of primary research on housing, various forms of quantitative and qualitative research come to mind. The distinction between quantitative and qualitative research is often taken for granted in social research. It is also the subject of much debate (Danermark et al. 2002: Chapter 6). Primary research on housing, however, extends wider than this distinction. So, at the outset, it is important to note that research, as the gathering of data relevant to housing, is often embedded within other genres where primary research is undertaken for another purpose. Indeed, most housing research appeals in some way or other to data as evidence in support of its argument. The focus here, however, is on housing research whose central role is primary research. Part of the task will be to distinguish primary research from these other purposes. While primary research can be embedded within different genres, the research component can be used by other authors for different purposes. Below I distinguish three genres that yield three different types of data: data gathered through the observation of events (observable data), data gathered on how participants experience events (participant data) and data which links events together and describes a particular process (descriptive data). I conclude by noting how historical data is included within primary research and not within history.

Observable data

The most prevalent type of data on housing is data which is produced by an observer counting events, either a whole population of events (a census) or a selected population of events (a sample). This type of data asks how often something occurs within a selected population of events.

Census data selects a population (persons, dwellings, households etc.) and within this population descriptively defines an event (such as characteristics of the population) and counts all instances of the occurrence of this event within a defined area at a defined time. A census is the complete enumeration of the population. Examples of census data on housing include data produced by national, international and regional statistical services such as the UK Office of National Statistics (2013), the United States Census Bureau (Mazur and Wilson 2011), the Australian Bureau of Statistics (ABS 2011a) and by social housing organisations (Scottish Housing Regulator 2013; UK Department of Communities and Local Government 2012; SCRGSP 2012; AIHW 2010).

For instance, the ABS Census selects a population of dwellings, descriptively defines a dwelling (ABS 2011a) and then counts the occurrence of dwellings within Australia. It further defines a range of events within this population such as tenure, type of landlord, type of dwelling, size of dwelling etc. and counts instances of these events. It thus provides a count of the distribution of dwellings and its characteristics at a particular time and place. This data provides a picture of a state of affairs at some past time. Further, because the counts of characteristics are related to a particular dwelling in a particular time and place, the frequency of counts in one time and place can be compared with another place at the same time or the same place at another time. They can also be combined in any number of ways (as cross-tabulations) to provide some detailed pictures of local areas at a particular time. As a result, census data show different patterns of the occurrence of events at different times and places.

Neuman (2006: Ch. 8) outlines various means by which samples can be determined: random sampling, systematic sampling, stratified sampling and cluster sampling. Similar to census data, the sample defines the population and the events to be counted at a specified time and location. Rather than all instances, only those within a selected sample of the population are counted.

So, the ABS Survey of Income and Housing Costs is a repeated cross-section survey of a random sample of households representative of the Australian population. At irregular intervals since 1982, the ABS has collected data on

> housing-related variables (housing tenure, dwelling structure and location, estimated house value, housing loans and repayments, housing costs etc.), labour market variables (wages, labour force position etc.), socio-demographic information (age, education, country of birth, family type etc.) and very detailed income data (specified by source of income and on a current weekly and previous financial year basis).
>
> (Dockery et al. 2008: 29; see also ABS 2009)

Sample data can be used not only to count events that are external to a person such as dwellings, rooms, households, tenures, activities, habits etc. but also events that are internal to a person such as decisions, motivations, views, attitudes, feelings, experiences, values, self-understandings/meanings, beliefs, opinions, expectations etc. While the range of some events admits of discrete differences, other events are continuous. Through the use of techniques such as a Likert scale, rank-order scale, Guttman scale etc., researchers demarcate segments along this continuum. In this way, levels of satisfaction, strength of desires, attitudes, preferences, feelings, beliefs etc. can be counted. For example, the National Social Housing Survey in Australia (Roy Morgan Research 2007) records the level of tenant satisfaction with services and dwellings, the extent to which social housing meets tenant needs and preferences.

Census and sample data specifies events in time and place. It can be one moment in time or over a specified period. It can be one place or it can be over a specified geographical area. Further, this data can be extended or extrapolated to estimate the occurrence of events at other places and/or times or, in the case of sample data, to estimate the occurrence of events for a whole population. For example, McNelis (2007) uses 2001 Australian Census data on low income persons over 55 years who are renting in Australia and the mortality rates of persons over 55 years to estimate future demand for social housing. These estimates are based on changes in data in the past which are projected forward to provide a state of affairs regarding the demand for social housing at some future time. Extrapolations in data to estimate future events make assumptions about the relationship between the past and the future about a certain population. The understanding sought here regards the likely occurrence of events in the future.

Participant data

Participant data is data on how a person experiences their situation. It is gathered through a variety of techniques: interviews, focus group discussions, personal reflections, personal observations about self and reactions/responses, and audio and video narratives. Through these techniques, the researcher seeks to 'take us, as readers, into the time and place of the observation so that we know what it was like to have been there. They capture and communicate someone else's experience of the world in his or her own words' (Patton 2002: 47). It describes 'lifeworlds 'from the inside out', from the point of view of the people who participate' (Flick et al. 2004: 3). Participant data includes decisions, motivations, views, attitudes, feelings, experiences, values, self-understandings/meanings, beliefs, opinions, habits, expectations etc. of a population. Three examples of such research are outlined below.

Hulse and Saugeres (2008) investigate home life, work and housing decisions and the linkages between government housing assistance and economic participation. A review of previous research evidence resulted in a series of questions on these linkages. To find answers to these questions, the study undertook in-depth interviews with 105 recipients of housing assistance. It reported on their 'attitudes, preferences and decisions' in relation to various types of economic participation:

> The research explores the factors that shape current decision-making about economic participation in the context of past experiences of housing and economic participation and aspirations and plans for the future. The research investigates both practical considerations for people in receipt of housing assistance in considering economic participation, such as balancing paid work with caring for children or maintaining mental and physical health, but also underlying social factors and cultural values about the role and value of different types of economic participation in relation to other contributors to individual and family wellbeing. (p.2)

With extensive quotes from interviewees, the study reported in some detail on various aspects of the linkage between housing assistance and economic participation. It collected data on the life histories of interviewees (family breakdown, residential mobility, child abuse, family violence, educational levels, employment and housing) along with their current attitudes to education, housing assistance and economic participation; their current view that 'paid work had both financial and non-financial benefits'; and their current views on paid work and barriers to entering paid work (mental and physical health problems, caring for children and other relatives, place/location, and transport and housing issues): 'Our qualitative research into the attitudes, preferences and decisions of housing assistance recipients has found that the factors that encourage or discourage them in making transitions into various forms of economic participation are complex and interrelated' (p.92).

A study of older persons in public housing (McNelis 2007; McNelis et al. 2008) sought 'to identify and explore more fully the policy and management issues confronting public housing providers' (McNelis et al. 2008: 3). It interviewed older people living in public housing, service-providers supporting older people, SHA frontline staff and SHA area/regional managers highlighting the differences in views expressed by these different groups.

A third and more complex example is research undertaken by Burke et al. (2007) who sought 'insight into how both renters and purchasers experience, identify and negotiate housing affordability in their everyday lives' (p.8) through a series of postal surveys, focus groups and in-depth interviews. In presenting their findings, the authors link certain demographic, income and housing characteristics with (i) affordability constraints and (ii) the observations, experiences, desires and perceptions of survey participants. The research goes further by presenting typologies of renters and purchasers. These aim 'to provide an effective means of capturing different ways of negotiating housing affordability by households through capturing particular drivers and similar outcomes' (p.56). Typologies are an imaginative synthesis that put together certain aspects of the data in different configurations, in this instance, a cluster of experiences, attitudes and strategies.

Descriptive data

Descriptive data describes a sequence of events in which only some events are understood by the participants. The description of events is pieced together by a researcher from the experiences of a range of participants either directly or indirectly. For instance, McNelis (McNelis & Burke 2004; McNelis 2006), in seeking to understand how social housing rent-setting and finance works in various countries, takes materials from various sources – documents, surveys, interviews, personal experience and accumulated knowledge – and describes the sequence of events in relation to rent-setting, operating finance and capital finance across a range of countries. These descriptions differ from country to country and within countries but they are then used as the basis for a more general theory of rental and finance systems (McNelis 2009a, 2009b).

Research by Hulse and Burke (Burke & Hulse 2003; Hulse & Burke 2005) illustrates the way in which housing researchers can descriptively put together material from various sources. As the study indicated in an introductory note:

> This was an exploratory study in view of the lack of any previous national, sector-wide research. Consequently, the research design involved use of multiple methods to build up an account of allocations systems in Australian social housing. These methods were both quantitative and qualitative, and the findings were compared and cross-checked against each other, consistent with the principles of triangulation. This approach recognised both deficiencies in available data and the likelihood of differing perspectives, given the complex and multi-layered nature of allocations, the inherent tensions of any form of administrative rationing system, and the dynamics of administrative systems dependent on social interactions for their implementation.
>
> Research methods included a documentary review of past social housing allocations policies and practices in Australia and a review of the literature. The research undertook a detailed analysis of available secondary data and scoped current policies and practices. This was supplemented by small-scale

surveys of housing practitioners in the public and community sectors, together with a policy workshop and interviews with key program managers which covered both formal and particularly informal allocations. The research also included a primarily web-based examination of overseas reforms.

(Hulse & Burke 2005: ii)

Descriptive data is not limited to economic processes but can also encompass linked but differing political and cultural meanings. For instance, Marston (2000) appeals to the differing views expressed in interviews and documents as evidence of changing discourse about public housing in Queensland and subsequent conflict between 'policy actors'; Blunt and Dowling (2006) appeal to a range of different views to show the different meanings of home; and Dodson (2007) appeals to discourses and practices around social housing in various countries, particularly Australia and New Zealand, to show how governments change housing policy through changes in discourse.

A note on historical data

Housing research is not limited to recent events. Observable data, participant data and descriptive data can be gathered, through various techniques, on the more or less remote past.

While techniques for researching the distant past and the current time may vary, both use similar methods. Granted, the task of historical research is much more difficult because (i) the data is more remote in time, (ii) the availability and extent of the relevant data is limited to already existing human expressions rather than what may be created by observing current events, undertaking interviews and discussions with current participants and putting together sequences of events as described by participants, (iii) the human expressions may be that of a culture quite remote from that of the historian, and (iv) gauging the status, accuracy and reliability of this data is more difficult.

Indeed, housing researchers today can call upon data that was gathered in the past for a particular purpose and use it for a different purpose. For example, the 1913 Joint Select Committee (of the Victorian Parliament) into Housing and the 1914–1918 (Victorian) Royal Commission into the Housing of the People in the Metropolis reported on overcrowding, poor standards, shortages, high rents and poor health of residents based upon interviews with witnesses and visits to 'slum areas' (Harris 1988; Howe 1988b). These reports, subsequently, became data for Harris and Howe as they sought to understand the history of social housing in Victoria. In doing so, they interpreted this data in a new context. For example, Harris noted 'the lack of statistical evidence on housing' and the complaint of one witness that 'available information dealt only with the whole of the metropolis', and concluded that 'most of the evidence taken during the Royal Commission tended to be impressionistic and value-laden'. Despite this lack of evidence, 'the inquiry…provided a focus for the housing reform movement and… for the expression of possible solutions to the housing problem' (1988: 10–11).

Housing researchers, whether investigating a particular topic today in various places or investigating the same topic in the near or far distant past, are still gathering data.

What are these primary researchers doing?

I began this review of primary research by distinguishing between (i) the data which the researcher turns to, attends to and gathers and (ii) the conclusion or viewpoint or significance they reach as an answer to a question. The review pointed to different types of data: observable data, participant data and descriptive data. This data can be gathered about current meaningful events or about meaningful events in the more or less remote past.

So what are these different primary researchers doing?

First, the field of materials that primary researchers turn to or pay attention to will be any expression of human understanding and living that is relevant to an understanding of housing. As noted in the introduction, these encompass both a range of external events that can be seen, heard, touched and smelt as well as a range of events that occur within each person. As noted in the epigraph for this chapter, these events are simply what are given. They are disparate, diffuse, unrelated and without significance. As events, they are unquestionable because they have already occurred. While a researcher can ask questions about their significance or about how they came to occur, as events they are prior to questions. That they occurred does not change as a result of questioning.[1]

Second, the field of materials attended to by the primary researcher and, subsequently, the type of data gathered differs according to the question asked. As questions vary, so too does the data to which they turn vary in answering that question: observable data, whether simple frequencies or more complex associations between events, answers the question: how often?; participant data answers the question: how do you understand what is happening?; descriptive data answers the question: what is the sequence of events that led to another event? So, the data turned to, attended to and gathered is a function of the question asked. In turning to and attending to the data, researchers are selective and their selection depends upon the originating question.

Third, researchers use different techniques to gather different types of data. Observable data, as the name suggests, pre-supposes definitions (usually descriptive definitions), uses surveys to count instances that meet these definitions and then collates and manipulates this data.[2] Participant data uses surveys, interviews, focus groups etc. to get participants to express their own understanding of themselves (motivations, feelings, values, beliefs) or perceptions of others. Descriptive data accumulates a picture or description of a sequence of events by supplementing knowledge gained from personal involvement with descriptions from other people involved at different points in the sequence.

Fourth, for all researchers, gathering data is a practical task. It demands a particular set of skills: the skill of observation; the skill of asking the right questions of people; the skill of structuring and organising surveys; the skill of collating and testing the validity of data; the skill of presenting data etc. It demands an attentive researcher, one with a capacity to notice details, albeit sometimes quite subtle ones. It demands the skill of conversation. It demands sensitivity to people and a capacity to relate directly and easily.[3]

Fifth, all the different types of data – observable data, participant data and descriptive data – document events that occur at a particular time (which can be aggregated into a period of time) and particular place (which can be aggregated into a specified geographical area). It is this concern with the occurrences of events in particular times and particular places that makes social science empirical. While Neuman, for instance, notes that different approaches to research tend to be associated with either quantitative or qualitative research, he also notes that all approaches are empirical: 'Each is rooted in the observable reality of sights, sounds, behaviours, situations, discussions, and actions of people. Research is never based on fabrication and imagination alone' (2006: 107). So, the primary question a researcher asks is whether or not certain events occurred.

Sixth, primary research not only documents a certain state of affairs at a particular time and place, it also seeks to associate the occurrence of events with one another either spatially, sequentially or both. It documents the extent to which, or the probability of, different events occurring simultaneously or in sequence, and the extent to which the same events occur in different geographical areas. Through techniques such as cross-tabulations, univariate analysis, bi-variate analysis, typologies etc. researchers can achieve some quite complex associations.

Seventh, primary research seeks to generalise its findings from one population at a particular time (or period of time) and particular place (or geographical area) to other populations in a different particular time and different particular place. For instance, a sample of a population may be generalised to the whole population or, considering some difference, can be adapted to another population in a different time and place. Thus, one of the key problems discussed in research is the conditions under which and the extent to which results can be generalised.[4]

Eighth, while some data is embedded within an interpretative framework that guides the research, such as the work of Hulse and Burke on allocations policy (Burke & Hulse 2003; Hulse & Burke 2005), most data, particularly observable data, is 'stand-alone'. The significance of this latter data is assumed. As Jacobs and Manzi (2000) note, this positivist approach continues to dominate housing research:

> the absence of *explicit* theory remains a defining characteristic of mainstream housing research…Within this paradigm, the task of the housing researcher is one of discovering objective facts, presenting them in a descriptive format in the expectation that policy makers will take notice and act accordingly.
>
> Though research within the empirical tradition achieves a level of sophistication in its analysis of social phenomena, its primary purposes are to establish facts and to prescribe effective action once problems are acknowledged. Not surprisingly, the conceptual categories used in housing research are rarely scrutinised within this paradigm; instead they rely upon the

collection of material evidence to reinforce policy recommendations. It would be erroneous to deny the benefits that ensue from such a view of housing research. Policy-oriented research enables academics to access resources and to ensure scholarship is up-to-date and close to the practical concerns of policy makers. However, there are disadvantages to such an integral connection with a practitioners' agenda. The resulting research product is often methodologically conservative. In addition, it is difficult to pursue new lines of investigation or, for that matter, to develop different conceptualisations of the policy process. Perhaps the most serious problem is that research of this kind is generally reactive to the professional housing lobby, which limits its opportunities to pursue a critical line of enquiry. Consequently, the positivist paradigm has had an impact on the *modus operandi* of housing research. Debates tend to be conducted within an agenda dominated by two competing ideologies: either policies should be formulated to bolster market mechanisms, or the role of the state should be extended. (p.35)

Later in the same article, in their advocacy of social constructionism, Jacobs and Manzi note the important relationship between the researcher and what is researched:

Social constructionism…offers an altogether different conception of reality from the one advanced by positivism, as well as a basis from which to understand the contexts and processes of housing…In particular such research emphasises the need to acknowledge both the importance of 'subjectivity' and how the act of research entails selection and pre-conceived idealisations, which, in turn, influence the research agenda. (p.36)

Primary research does not take place in a vacuum. It is not presupposition-less. It entails 'selection' and 'pre-conceived idealisations'. Insofar as a researcher is answering a particular question, it entails a selection of events. The process of selecting is also a process of excluding. This selection and exclusion is a function of the question asked. This question regards the significance of some selected event or events to the exclusion of other events. Neuman (2006), De Vaus (2002), Crotty (1998) and Blaikie (2003) rightly argue that different approaches are associated with different traditions in social theory, different research techniques and different sets of data.

As a result, the events attended to, focused on and considered relevant depend upon the approach of the researcher: an interpretative (or social constructionist) approach will pay attention to the way in which each person constructs their world; a critical approach will pay attention to the way in which power is exercised within a society; a feminist approach to housing will pay attention to the experiences and observations of women which are overlooked by non-feminist approaches, etc.

Finally, just as research itself is not presupposition-less, so too the expressions of human understanding and living that are turned to and gathered by researchers have their own pre-suppositions. Each person expresses themselves in words and actions, yet these words and actions incorporate human intelligence and stupidity, responsibility and irresponsibility, collaboration and alienation, love and hatred, self-sacrifice and self-aggrandisement of persons, groups and nations. Whether a person can adequately express the motivations, beliefs, feelings, understandings etc. has its conditions: it depends upon their language skills, their capacity to put together the right words and phrases; it depends upon their competencies, their ability to put together a series of activities that together create what they want – to build a house requires competencies in design, soil assessment, building materials, construction etc. As a consequence, for researchers, the expressions of human understanding and living, the data, are ambivalent and their significance is yet to be determined. This significance will depend upon the extent to which these expressions make sense, are intelligible and are responsible.

It is to this complex question of presuppositions that we now turn to develop a more precise understanding of Research.

Towards a more precise understanding of Research

Housing research has sought answers to a range of questions. Through its sophisticated techniques it has gathered a vast array of data and has contributed to an understanding of housing. But do its techniques and achievements qualify it as a scientific approach? I suggest that they do not. This, however, raises some difficult questions about presuppositions that are not so much questions for Research but rather for Foundations. A discussion of these will be left until Chapters 7 and 8. But some indication of the presuppositions of Research is necessary for a more precise understanding of Research.

A first presupposition regards distinguishing common-sense understandings from theoretical understanding. Neuman (2006) identifies five competing approaches to research – positivist, interpretative, critical, feminist and postmodern – each of which produces different results. In my assessment, however, these five competing approaches are not scientific approaches to housing. The dominant orientation of all approaches is immediately practical. They view things in terms of their immediate practical value. It is within the everyday taken-for-granted world of common sense that we live, develop and discover meaning. It is within this world that we develop our interests and commitment to what is of value. It is in view of the demands of everyday living that we attend to and select data, we deem some data more significant than other data. Indeed, we ignore or refuse to acknowledge other events because they are beyond the scope of our interests.

As C. Wright Mills notes in *The Sociological Imagination*: 'the everyday empiricism of common sense is filled with assumptions and stereotypes of one or another particular society; for common sense determines what is seen and how it is to be explained' (1970: 138). The positivist approach presupposes the dominant culture within a society. By assuming this context, data can 'stand alone'. Besides this dominant culture within a society, there are also other non-dominant or oppressed cultures. Researchers from other traditions can also operate within the parameters of these cultures and direct their attention towards what is significant within that particular (minority) culture. These researchers subtly and unknowingly give priority to some things over other things, some data is deemed important to collect whereas other data is deemed irrelevant.

Neuman describes the five approaches as competing approaches. However, it should be noted that (i) groups (and a common culture) can form around any number of interests and so it may be possible to discern any number of approaches, (ii) as people shift between groups, they 'spontaneously' adapt to its culture and (iii) depending on the 'story' they want to tell, housing researchers can shift between various approaches to housing research.

While housing researchers using these other approaches speak of their approach as theoretical and have an awareness of the role of theory as a heuristic within research, they have not thrown off the vestiges of the everyday taken-for-granted world of common-sense living and have an inadequate notion of theory. Each approach is inadequate as a scientific approach because it does not make the radical and decisive shift to the world of theory. Of course, this claim hinges on an understanding of what is and what is not theory. I will return to this issue in Chapter 5 and Chapter 7.

A second presupposition is that Research as a functional specialty has a precise technical role or function as one of eight functional specialties and it is within this larger context that the role of primary research needs to be understood.

I have already pointed to the two movements within asking a question – the first movement turns and attends to the data; the second is a discovery or an answer. The initial view of Research, however, did not make any distinction between different types of questions. I simply sought to note how, spontaneously, a question distinguished between data which is attended to and the meaning or significance of that data reached in the answer. As we have seen in Chapter 3, within an everyday common-sense framework, a housing researcher asks all sorts of questions in different ways. In asking these questions, they attend to different data; in reaching an answer, they select different data. The data attended to and the answers reached depend upon their interests and cultural presuppositions.

For a scientific approach to housing, however, it is important to attend to and distinguish different types of questions. Further, it is important to acknowledge that one type of question, a what-is-it-question, precedes other questions. Before we can ask how often something occurs, how it changes over time (through history), what its value is and how we can improve it, we need some precision as to what it is. Later, in Chapter 5 on the functional specialty Interpretation, I will argue that a theory of housing is an answer to the question 'what is housing?', an answer which seeks an explanatory definition of housing. It is by answering this question that we can distinguish 'housing' from other things. It is within this context, the context of theory as an answer to a what-is-it-question, that the functional specialty Research has a specific role or function which is to gather and make available the data that is relevant to answering a what-is-it-question.

A third presupposition is that science builds upon its past achievements. Thus, Research will presuppose some already developed theory. In its relation to Research, a theory of housing will function as a heuristic, an anticipation of the significant and relevant data. So a Researcher does not begin with a blank mind (or *tabula rasa*), but rather with the best heuristic or the most up-to-date theory. It is a theory which gives the best current available answer to the question: what is housing? As a heuristic it guides future Research. Again in Chapter 5 I will come back to this issue.

A fourth presupposition is a stance which recognises that every answer to the question 'what is housing?' is incomplete. While a theory as a heuristic guides primary research, Researchers are not simply passive recipients of the latest theory. As the epigraph to Chapter 3 noted, science is the dynamic interplay between the relevant data and theory. So, the functional specialty Interpretation contributes heuristics that anticipate the significance of a set of events. Research, however, as it is governed by the original question 'what is housing?', seeks to go beyond a particular answer to the question. It does so by continuing to ask further questions about the data pointed to by this heuristic. Insofar as Researchers notice inconsistencies, unaccounted for events, something that has been missed, there is the possibility that they can point out the inadequacies of the current theory, and the need for a new context or perspective for understanding the significance of the data, the need for a new theory. So, within Research, we can distinguish between standard and non-standard research. Standard research is concerned with affirming a current theory at a different time and place or within a different field of study. But non-standard research seeks to notice inconsistencies and events unaccounted for that will upset the current theory. The role of Research, then, is not simply about moving with the stream of standard research but, more fundamentally, about upsetting the stream, about provoking the Interpreter to incorporate these events into a more adequate theory.

Philip McShane (2007a: 52) puts it this way:

> Research, then, seeks to notice the unsettling, the anomalous from the viewpoint of the present *nomos*...The viewpoint of the present *nomos*, in so far as you are interested in participating in, promoting, advocating, functional specialist work, is your probably quite inadequate viewpoint...That is what you bring to the present research...Research, as a functional specialty, is asked to do this, and no more: no more, that is, except the positive functional move of handing on the problem to the theoreticians, who have to somehow place the new plant, particle, perspective. A tough job.

Implications for housing research

Housing research seeks answers to questions. Within a question we can distinguish two movements: a first movement turns to the occurrence of events which provoked the question and about which we inquire; the second movement is the discovery of an answer to the question.

My review of housing research illustrated how this distinction is implicit. In seeking answers to questions, researchers turn to the occurrence of events within the field of materials. These events are the data. They occur prior to the researcher's question. In this sense, they are unquestionable. They remain the same no matter what the result of the questioning may be. They are disparate and diffuse and without significance. It is this appeal to the occurrence of events that constitutes a science as empirical. In seeking an answer to their question, researchers move back and forth – on the one hand, towards the data; on the other hand, towards a discovery, an answer to their question. It is in discovery that disparate and diffuse data are related to one another and the significance of the data (the events) emerges.

The review also showed how approaches asked different questions which led researchers to attend to and select different data. Moreover, the researcher is not simply satisfied with grouping, defining and categorising events, they also seek to associate events with other events either spatially, sequentially or both. But such associations are relationships in time and place. They are a necessary condition for understanding the significance of events, but are not sufficient for such understanding.

Research asks the empirical question: what events are occurring in this time and place and to what extent are these events associated? An empirical question anticipates the occurrence of events in time and place: it anticipates answers such as the frequency with which events or properties or characteristics occur; it anticipates answers such as participants' description of feelings, beliefs, attitudes, understandings etc.; it anticipates that events occurs in a sequence. As a result, Research yields observable data which shows the extent to which certain elements in housing are associated (in time and place) with one another; Research yields participant data which shows the extent to which certain elements in housing are associated because participants use housing for their own purposes in particular ways; and Research yields descriptions of sequences of events. The answers to an empirical question can be aggregated into a population of events within a specified time-period, within a specified geographical area or both.

Research, however, is not without suppositions. The researcher operates in a context from which particular questions emerge. The positivist operates in the context of the everyday taken-for-granted world of common-sense living and presupposes the particular culture dominant within a particular society. Other approaches such as interpretative, critical, feminist and postmodern also operate in this common-sense world but presuppose a particular non-dominant culture within that particular society. The key problem is not so much that the data turned to, attended to and gathered is a function of a particular approach to social research but rather the adequacy of an approach that operates within the everyday world of common-sense living.

As a functional specialty, Research operates in the context of Functional Collaboration as a whole. Its role presupposes the world of theory. It presupposes distinctions between different types of questions and recognises that a what-is-it-question precedes other questions. As we will see in the next chapter, an answer to the question 'what is housing?' is a theory of housing. Research presupposes some answer to this what-is-it-question, however inadequate it may be. The role of Research is to notice the anomalies, the inconsistencies and the unaccounted for data in the current theory and to push the Interpreter or theorist beyond the current theory of housing towards a more adequate theory. Its role is to search for the occurrence of events which are not incorporated into the current theory of housing.

Such a view of the functional specialty Research has profound implications for current housing research. Operating within the taken-for-granted everyday world of common-sense living, current housing researchers ask any number of questions and seek answers to these. Their orientation, however, depends upon the researcher's location within their particular common-sense world with its particular interests. While they ask any number of different types of questions and seek answers to them, their common-sense orientation limits the scope of the data they regard as significant. While their answers address the immediate practical problem, they are inadequate.

A scientific approach to housing, however, recognises that empirical data is related to and contributes to one particular question, a what-is-it-question. Research gathers data or evidence relevant to answering this question. The functional specialty Research, then, envisages a paradigm shift in primary housing research *from* answering any type of question raised *to* answering an empirical question, what events are occurring in this time and place and to what extent are these events associated, within the context of answering the theoretical question: what is housing?

But this more precise understanding of Research presupposes a more precise understanding of theory. It is to this issue that we turn in the next chapter on the functional specialty Interpretation.

Notes

1   For a more detailed discussion of 'the given', see Lonergan ([1957] 1992, in particular 405–407).
2   Kemeny discusses some of the difficulties associated with defining what he calls 'first order' concepts in housing studies such as 'dwelling' and 'household'. As noted in Chapter 5, this problem arises where a descriptive definition is used rather than an explanatory definition (1992: 21ff).
3   This set of skills is different from the set required of a theorist: the skill of asking further questions; the skill of shifting contexts; the skill of picking from a range of possibilities the one that best relates the data. The theorist requires a capacity to grasp possibilities and to work them into new arrangements.
4   See, for example, Blaikie's discussion of inferential statistics (2003: Chapter 6).

5 Interpretation and the theoretical question

Dialectic   Foundations

History   Policies

Interpretation   Systematics

Research   Communications

An interpretation is the expression of the meaning of another expression. It may be literary or scientific. A literary interpretation offers the images and associations from which a reader can reach the insights and form the judgments that the interpreter believes to correspond to the content of the original expression. A scientific interpretation is concerned to formulate the relevant insights and judgments, and to do so in a manner consonant with scientific collaboration and scientific control.

Bernard Lonergan in *Insight: A Study of Human Understanding*
([1957] 1992: 608)

Since Jim Kemeny wrote *Housing and Social Theory* in 1992, the role of theory within housing research has become much more prominent. Kemeny sought to further the theoretical development of housing research (O'Neill 2008) and outlined two requirements: first, 'the need consciously to re-integrate housing into broader issues of social structure', and second, 'to develop a greater sense of reflexivity, by examining more closely the epistemological grounds of what we are doing' (Kemeny 1992: xvii).

Kemeny argues that each discipline in the social sciences (sociology, economics, psychology, politics and geography) 'dimensions out' or abstracts different aspects of social relationships. Each has its own set of conceptual tools, its own mode of discourse and its own debates. He puts it as follows:

> In general, the social science disciplines can be seen to be based on dividing the social world into a number of dimensions. Sociology, for example, 'dimensions out' social relationships which are often conceptualised in terms of the abstraction known as social structure. Economics does the same for the market. Psychology dimensions out individual mental processes. Political

science dimensions out power and political institutions. Geography dimensions out space; history dimensions out time, and so on.

Each discipline develops its own sets of conceptual tools for the analysis of its particular dimension. Theories are explicated and tested, and a characteristic mode of discourse is evolved through the generations, with its own major debates and controversies. The point about this is that each discipline is based on researchers being 'disciplined' into thinking in certain ways and in critically evaluating existing theories and concepts developed by others within that mode of discourse.

Disciplines are based on a process of conceptual abstraction. That abstraction provides the epistemological basis for the discipline and provides it with a selective frame of analysis. Disciplines are not normally defined in terms of a concrete field or subject of analysis...They are more usually defined by a frame of reference, even if some frames of reference prove in practice to be more amenable to theorising than others.

(Kemeny 1992: 3–4)

While there is a common reference point for each discipline – the social world – Kemeny does not address the issue of how these different disciplines are related to one another. Without some heuristic that does so, it is difficult to see how housing research can be truly interdisciplinary or transdisciplinary (Lawrence 2004; Lawrence & Després 2004).

Kemeny addresses the second requirement by raising questions about 'the substantive focus of housing research, the methods we use, the data sources we tend to take for granted, the questions we pose and the received wisdoms, often from policy-makers, about what does and does not constitute 'a housing problem'' (1992: xvii). He examines and critiques 'first-order concepts' such as 'household' and 'dwelling', raising questions about how we use and define them within housing research; he notes the lack of debate in housing research about quantitative and qualitative research, how statistics are socially constructed, and questions the reliability of official statistics; he draws attention to how housing policy debates interweave data, methodology and concepts, particularly in the development of 'higher level composite concepts' such as 'overcrowding', 'dwelling type', 'housing subsidies', 'forms of tenure', housing class and structuration; and he highlights the importance of researchers reformulating housing problems from a theoretical perspective rather than allowing particular interest groups to define the problem in their own way (1992: 20f).

In a series of articles in a 2009 special issue of *Housing, Theory and Society* celebrating twenty-five years of the journal, David Clapham (2009), Peter King (2009) and Chris Allen (2009) discussed the ongoing debate on the role of theory in housing research while Bo Bengtsson (2009a) and Kenneth Gibb (2009) sought to show the value of theory to housing research from their respective disciplines of political science and economics. As Clapham noted in his introduction, the articles spanned a range of views on theory and housing research: Bengtsson and Gibb implicitly supporting the view of Kemeny and others that there is no such thing as 'housing theory' but rather social theories that can be applied to housing; King arguing for a theory of housing; and Allen arguing that housing research has no claim to special status among differing views of housing.

Two different approaches to theory – whether housing researchers should simply *apply* social theories from various disciplines or should work to *create* housing theory – imply different understandings of theory. If we are to come to a position on this issue, we need, as King suggests, to answer a further question: what is theory?

Interpretation: an initial view

The previous chapter used a phenomenology of research to distinguish two movements in asking a question: the first movement turns to the events or the data which provokes the question and of which we seek an understanding; the second movement is the discovery of meaning or significance in data and is an answer to the question. This discovery is what my initial question intends. It is arrived at only at the end of my inquiry. It integrates or provides a higher viewpoint on some set of data within the field of materials. In an initial way this phenomenology of the question distinguishes the functional specialty Research and the functional specialty Interpretation.

Further, I noted that the functional specialty Interpretation provided the context for Research: the role of Research is to provide the data relevant to answering a what-is-it-question; the role of Interpretation is to answer that what-is-it-question. So, while the functional specialty Research is concerned with occurrences of 'housing' that precede our questions, the functional specialty Interpretation is concerned with specifying just what this 'housing' is. Interpretation is concerned with defining just what the object of our research is.

Defining what is being researched

A what-is-it-question precedes other questions in research. It can be distinguished from other types of questions: how-often-questions, change-questions, is-it-worthwhile-questions and what-to-do-questions. As I will show later, these are the concern of other functional specialties. But it is also important to note the difficulty I face, here, in reviewing the research on housing. The distinctions between these different types of questions are often not made within one book or article but are conflated or run together and assumed without rigorously distinguishing between them. So articles/books will discuss how housing (or the language of housing) is used to aggrandise some particular person or group and how it is made to serve their interests without clearly defining what housing is, or will discuss bringing about a new state of affairs within housing (or some new housing policy) without considering what it is and whether it is worthwhile promoting or creating.

Housing research deals with the definitional question in three principal ways: by defining terms either formally or informally, by outlining a 'conceptual framework' or an 'analytical framework' that will inform the research, or by using participant descriptions.

Defining terms

The question of defining terms arises in different settings. When gathering statistical data, terms are defined through the use of data dictionaries or glossaries. These definitions allow data to be allocated as defined. For example, *Housing Statistics in the European Union 2010* (Dol & Haffner 2010) includes an appendix that sets out the definitions used by Eurostat and different European countries for dwelling, household, overcrowding, social versus private rental dwelling etc. The ABS Census Dictionary (2011b) defines terms such as dwelling, household, landlord, tenure, rent etc.

A more complex example is the way in which performance indicators for social housing are defined by 'putting together' different data to form a particular performance indicator. For instance, under the UK National Affordable Housing Program, housing quality is defined and assessed through the combination of ten indicators each of which is 'made up' of various components (UK Housing Corporation 2008). AIHW (2010: 24) defines the performance indicator, affordability, in terms of (i) the average weekly rental subsidy per rebated household and (ii) the proportion of rebated households spending not more than 30 per cent of their income in rent. In turn these components are defined as combinations of further components.

Conceptual frameworks

Conceptual or analytical frameworks seek to bring together the different aspects of the object under inquiry. They highlight and connect together key terms associated with the object in a way which gives coherence to a discussion of it. Below seven conceptual frameworks are discussed as they have been used within housing research: four housing provision frameworks – three from an economic perspective and one from a political perspective; and three frameworks for elements within housing provision – rental systems, allocations systems and the structure of a house.

Housing provision: an economic perspective

Over the past three decades or more, various analytic frameworks for housing provision have been proposed and subject to review and analysis (for example, Kemeny & Lowe 1998; Lawson 2006: Ch. 4). Here, however, only a couple of these are discussed as illustrative of analytical frameworks for housing provision.

Peter Ambrose (1991) proposed the Housing Provision Chain model as an analytic framework for comparing housing provision systems. He distinguished five stages as a linked sequence of events: promotion, investment, construction, allocation and management. He further distinguished between the activities of 'non-democratically accountable' actors or agencies (private sector) and of 'democratically accountable' actors or agencies (public sector). This framework allowed him to distinguish between different types of housing policies and subsidies according to each stage of the chain. After a description of housing provision in three areas – Berkshire (Britain), Toulouse (France) and the E4 corridor of Stockholm (Sweden) – and working out a set of objectives for housing policy, he uses the Housing Provision Chain as a way of highlighting aspects of each housing provision system and to show how each could achieve their objectives (Ambrose 1992). Sasha Tsenkova adapts the Housing Provision Chain model for her own purposes, conflating the first two stage – promotion and investment – into one, and spelling out in more detail the public sector and private sector agencies involved in housing provision. She uses it to discuss the shift from predominantly public sector to private housing provision in Bulgaria (1996).

Drawing on the work of Ball and Harloe (Ball 1986; Ball et al. 1988; Ball & Harloe 1992; Ball 1998), Terry Burke (1993; Burke & Hulse 2010) compared housing provision in different countries. Ball et al. (1988: 5) describe their analytic framework, 'structures of housing provision' thesis, as follows: 'A structure of housing provision encompasses the inter-relations between all the agencies involved in the production, exchange, finance and consumption of housing in a particular way.'

In *International Low Income Housing Systems* (1993), Burke uses the 'structures of housing provision' thesis to propose a framework of analysis consisting of four sub-systems and a systems context as a way of understanding and comparing housing systems.

> The four sub-systems are those of *production* concerned with the nature and techniques of land ownership, land assembly and housing production, of *consumption* concerned with the forms and methods by which people and households consume housing, of *exchange* concerned with the practices and institutions which facilitate the exchange of housing, i.e., finance, and *management*, that is the practices by which the housing stock is managed including planning.
>
> The 'systems context' refers to the economic, demographic, administrative, legal, and political processes which shape the form, timing and direction of housing policy relating to each of the four sub-systems.
>
> (Burke 1993: 14)

Rather than beginning with the structures of housing provision, Burke begins by developing a typology of contexts within which housing systems operate. He recognises the importance of this context for the way in which the four sub-systems of housing provision are structured. It is through this overall context that different countries manage the diverse range of economic, social and environmental demands. In his typology of contexts, Burke identifies three broad approaches to societal management: market liberal/economic rationalist, welfare corporatist and liberal corporatist (a mixture of the other two). He is then able to compare the housing implications of each approach by focusing his description and analysis on the four sub-systems of production, consumption, exchange and management. For Burke, an understanding of housing provision is defined in terms of these four sub-systems.

In her book *Critical Realism and Housing Research*, Julie Lawson (2006) seeks to explain why the housing solutions in The Netherlands and Australia have diverged as they evolved since the late nineteenth century up to the end of the twentieth century. In presenting her explanation, Lawson moulds together four different questions regarding (i) her starting point in critical realism, (ii) the subject that she is researching, viz. housing provision, (iii) the emergence of particular housing solutions or forms of housing provision in Australia and The Netherlands and (iv) an explanation for divergent housing solutions or different forms of housing provision in the two countries.

In my view, these are questions for different functional specialties. The third question concerning the emergence of particular housing solutions would be considered in the functional specialty History, whereas the first question on her methodological starting point and the fourth question which compares housing solutions in Australia and The Netherlands would be considered in the functional specialty Dialectic. My interest here is with the second question: how Lawson defines the subject of her research, what it is she is seeking to explain.

Lawson is seeking to explain divergent housing solutions. Consistent with the critical realist perspective, each solution emerges from a cluster of causal mechanisms, the necessary and contingent relations. These two sets of relations operate together to produce an outcome:

> Necessary relations are not fixed behavioural laws or predictors of events. They do not exist as isolated atoms in a laboratory. There are no standard definitions of necessary relations applicable to all time and space…Necessary relations are actualised in the context of other sets of interacting contingent relations. For this reason, concrete historical case study research is an integral part of the explanatory process. (p.28)

Lawson defines housing provision in terms of three necessary relations between agents: property relations, investment and savings relations, and labour and welfare relations. But these cannot be defined universally but rather are defined within the context of contingent relations. These contingent relations change how the necessary relations are actualised in a particular situation.

> The term housing solution refers to the coherent fit between social relations underpinning a housing system and the practical solutions and outcomes produced. It is contended that housing solutions in Australia and The Netherlands have emerged from the fundamentally different packaging of property, investment and savings, and labour and welfare relations, which have promoted distinctive housing choices and living environments. Most Australian households aspire to home ownership and reside in large, low-density cities. In The Netherlands, until recent years, social rental housing

has been the dominant tenure in relatively numerous compact towns and cities. (p.24)

By defining housing provision in terms of the necessary relations of property relations, investment and savings relations, and labour and welfare relations within the context of the contingent relations of risk, trust and state, Lawson can investigate the actual emergence of divergent housing solutions in Australia and the Netherlands.

Housing provision: a political perspective

In two reviews of housing studies, Bo Bengtsson (2009a, 2009b) notes that housing researchers seldom 'employ a *politics perspective*, analysing the political institutions of relevance to housing provision and the games and processes of decision-making per se' (2009b: 1).

Lennart Lundqvist was one political scientist who did develop an analytic framework of housing provision from a political perspective. In *Dislodging the Welfare State? Housing and Privatization in Four European Nations* (1992), he outlines a framework for understanding privatisation of housing in Great Britain, The Netherlands, Norway and Sweden. Lundqvist does this in three stages. First, he examines the different meanings of privatisation (p.3) and subsequently defines privatisation as 'actions taken by actors legitimately representing the public sector to transfer the hitherto public responsibility for a certain activity away from the public and into the private sector' (p.3 quoting from Lundqvist 1988: 12). It is this relocation of responsibility that is 'the main criterion for delimiting 'privatization' in a general policy sense' (p.3). Second, he defines the field within which he is investigating 'privatization'. He views the housing sector as a system that is continually adjusting households and dwellings to each other. Two key factors can be singled out: household purchasing power (consumption) and dwelling price (production). In this way, he develops a definition of housing policy that is not limited to particular courses of action being pursued by governments but rather from 'the necessary and general logic of housing provision'. This definition allows for a comparative analysis which includes not only the elements present within a country but also of 'why this or that element is not present in a policy or why particular patterns of public-private responsibilities in the housing sectors of different countries occur, develop, and change' (p.4). Third, within the processes of consumption and production Lundqvist distinguishes the economic aspects of housing provision from the political aspects; those 'particular courses of action or inaction' open to any government (outlined diagrammatically in his Figure 1.1). From three forms of government intervention – regulation, financing and production – he develops a taxonomy of privatisation policy alternatives (outlined diagrammatically in his Figure 1.2).

Having established this analytic framework, Lundqvist then goes on to postulate that privatisation can be explained in terms of a modified 'power resources' approach (as proposed by Walter Korpi (1989)) but within an institutional context (pp.7–8). Three dimensions of this institutional context are particularly relevant to an explanation of privatisation in the four countries he studied: the legacy of earlier policy whether supplementary or comprehensive, the strength of public bureaucracies and the organisation of affected interests (pp.9–10, 122).

Elements within housing provision

RENTAL SYSTEMS

McNelis and Burke (2004) and McNelis (2006) in two related AHURI papers develop a framework for understanding social housing rent-setting. In the first paper, McNelis and Burke describe how rents are set in different social housing systems in Australia and overseas as well as the problems emerging within Australia. In the second paper, McNelis outlines a framework for understanding the two basic types of rental systems and variations within them. The first basic type is property rental systems such as a cost-rent system, a market-rent system and a market-derived-rent system. The second basic type is household rental systems such as an income-related rental system, a subsidy rental system and a flat-rent rental system. He describes the principles that underlie and distinguish the various rental systems. But this description of the differing processes for setting rents raises a series of questions as to the appropriateness of particular rental systems for social housing:

> The discussion about various rental systems and their underlying principles and processes raises a further question. Which rental system is more appropriate for social housing and the achievement of its objective? This raises a question about the relationship between rental systems and the objective of social housing.
>
> (McNelis 2006: 41)

McNelis goes on to locate these different rental systems within a variety of social housing finance systems and four objectives of social housing often associated with rental policy – affordability, equity, workforce incentives and the autonomy of SHAs.

These papers thus provide a framework for understanding particular operating rental systems and their relationship with social housing objectives.

ALLOCATIONS SYSTEMS

In their AHURI report on allocations systems in social housing, Kath Hulse and Terry Burke (2005) provide a national overview; examine the pressures on these systems, the responses of providers and emerging concerns; explore the variations in practices as well as the perspectives of housing workers; review initiatives in Australia and overseas; and develop a framework to facilitate learning and decision-making. Prior to this, they provide 'a framework for considering allocations…to enable policy makers and others to conceptualise allocations systems and their linkages with other aspects of social housing operations' (p.3). In developing their conceptual framework, Hulse and Burke explore four aspects of social housing allocations.

First, they compare the similarities and differences between private rental and social housing. At first view these allocations systems appear very different, but a more considered view shows that these differences are more a matter of emphasis. For instance, at first view, private rental provides individual household choice while social housing is dominated by administrative criteria and process. But a more considered view shows that both to varying degrees incorporate trade-offs and household choice as well as administrative criteria. In private rental the primary factor is 'ability to pay market prices' while in social housing the primary factor is 'housing need'. Both models incorporate specific factors such as choice of provider, information, application, eligibility assessment, household choice, order of access to housing, matching households and properties, and consideration of neighbourhood impact. But for each factor, each model has different characteristics (pp.4–5).

Second, they explore social housing allocations as administrative rationing and define the system as a 'multi-layered process' in which there are three layers: a strategic planning process which determines the agency's broad role in social housing; primary rationing which defines which groups of households are eligible; and secondary rationing which matches individual households with individual properties (pp.6–7).

Third, they distinguish between (i) the formal rationing process incorporated in documents and (ii) the informal rationing process, the discretionary interpretation of policies and guidelines made by housing workers.

Finally, they note the linkages between allocations and other aspects of social housing, in particular, how allocations influence and are influenced by rent-setting and by tenancy and property management. In doing so, they relate each layer of the allocation process to these other two aspects of social housing.

Now we can note some characteristics of this conceptual framework for social housing allocations. Firstly, this framework is something new within housing research which is still in its early days of developing such frameworks. Secondly, the framework defines the elements that are considered in the remainder of the report, thus providing a particular focus for this research. For instance, Chapter 5 is developed around the three layers: strategic planning, primary rationing and secondary rationing. Thirdly, the framework comes at the beginning of the report and as such gives the impression that it precedes their research rather than being the result of research. Fourthly, the framework does not work to integrate access in private rental and access in social housing, preferring to describe the characteristics of social housing allocations. Finally, while a number of characteristics are outlined, they are not systematically related. Rather, the characteristics of the conceptual framework separately describe the allocation process in both private rental and social housing without moving to an understanding of what is systematically core to both forms of tenure.

THE STRUCTURE OF A HOUSE

One element of housing provision is its design. In a *Housing Studies* article, Kim Dovey (1992: 177) 'explores the meaning of the house through an interpretation of model house advertisements and display houses in Australia over the past 20 years'. His interest is in the way in which house plans have changed over time and reflect changing relationships within the house (children and parents, and gender) and the changing relationship between the house and the world outside (status and identity).

To do this, Dovey defines the structure of the house (or, in his terms, uses 'a technique') as

> composed of a range of 'places' or centres of meaning which are structured in a certain set of relationships. These 'places' are meaningful segments of domestic space which may or may not be enclosed. The plan signifies a set of categories and their structural relationships. (p.178)

This definition allows him to focus on these 'meaningful segments of domestic space', showing how their size, function and relative location have changed and interpreting the meaning of these changes in terms of internal and external relationships.

Participant descriptions: the meaning of home

Housing is not just about the physical structure of houses and housing provision, nor is it simply about tenure or housing policy. It is also about homes: 'sites of emotional, cultural and social significance' (Dowling & Mee 2007: 161); 'a *spatial imaginary*: a set of intersecting and variable ideas and feelings, which are related to context, and which construct places, extend across spaces and scales, and connect places' (Blunt & Dowling 2006: 2).

The research around the meaning of home highlights the varying definitions (for example, Kemeny 1983; Johnson 1996; Corrigan 1997: Ch 7; Easthope 2004; Lloyd & Johnson 2004; Mallett 2004; Blunt & Dowling 2006; Dowling & Mee 2007). The difficulty of defining home is such that most authors recognise that home is a 'multidimensional concept' and the research provides understandings of home as expressed through popular magazines (for example, Lloyd & Johnson 2004), life stories (diaries, memoirs, autobiographies, interviews, focus group discussions etc.), novels, magazines and household guides, images (art and film) and designs (for example, Blunt & Dowling 2006). These varying sources provide understandings from a very broad range of people: residents, owners, researchers, designers, philosophers etc.

After critically reviewing a range of understandings of home including understandings from research and understandings developed by researchers, Shelley Mallett (2004) concludes with three pertinent questions: 'How then is home understood? How should home be understood? Or, how could home be understood?' (p.84).

She continues:

> Clearly the term home functions as a repository for complex, inter-related and at times contradictory socio-cultural ideas about people's relationship with one another, especially family, and with places, spaces, and things. It can be a dwelling place or a lived space of interaction between people, places, things; or perhaps both. The boundaries of home can be permeable and/or impermeable. Home can be singular and/or plural, alienable and/or inalienable, fixed and stable and/or mobile and changing. It can be associated with feelings of comfort, ease intimacy, relaxation and security and/or oppression, tyranny and persecution. It can or can not be associated with family. Home can be an expression of one's (possibly fluid) identity and sense of self and/or one's body might be home to the self. It can constitute belonging and/or create a sense of marginalisation and estrangement. Home can be given and/or made, familiar and/or strange, an atmosphere and/or an activity, a relevant and/or irrelevant concept. It can be fundamental and/or extraneous to existence. Home can be an ideological construct and/or an experience of being in the world. It can be a crucial site for examining relations of production and consumption, globalisation and nationalism, citizenship and human rights, and the role of government and governmentality. Equally it can provide a context for analysing ideas and practices about intimacy, family, kinship, gender, ethnicity, class, age and sexuality. Such ideas can be inflected in domestic architecture and interior and urban design. (p.84)

Rather than working towards some sort of definition of home, herself, Mallett prefers to maintain the multidimensional view of home and, referring to Hollander (1991), argues that 'both the meaning and study of home 'all depends''. She concludes: 'how home is and has been defined at any given time depends upon 'specification of locus and extent' and the broader historical and social context' (p.84).

What are these researchers doing?

The primary focus of these three genres of research is on defining an area of research: the first focuses upon defining the way particular terms are used, their scope and their limits; the second, the conceptual framework, is a more complex definition that guides research in a particular area; and the third points to participant descriptions.

So what are these researchers doing when they are defining an area of research?

First, in the previous chapter on the functional specialty Research I noted that 'the *starting* point of social science is to be found in ordinary social life' (Schutz 1972: 141). This ordinary social life is a world of meaningful events that occur at particular times and in particular places. In developing their definitions, these researchers are paying attention to these meaningful events. For example, in defining the structure of the house, Dovey points to 'places' or 'centres of meaning which are structured in a certain set of relationships'; participant descriptions pick up everyday meaning of terms.

Second, these researchers are defining housing (or some aspect of it) in terms of other events with which it is regularly associated. These events are associated in time and place and regularly occur at the same time and/or precede or follow on in some pattern. The definition hinges upon an understanding that grasps the meaning of apparently disparate meaningful events in a single perspective and thus stand in relationship of the higher to the lower where the higher is the single perspective and the lower are the individual meaningful events. For example, Lawson defines housing provision in terms of property relations, investment and savings relations, and labour and welfare relations. An understanding of 'housing provision' grasps these three relations in a single perspective.

A third aspect of definition is illustrated in Burke's use of the structures of housing provision (SHP). Burke superimposes upon SHP what he calls a 'systems context'. He uses SHP but is aware that it operates in a larger context of 'economic, demographic, administrative, legal, and political processes' (1993: 14; see also Burke & Hulse 2010). This context shapes the form of SHP in each country and it is this higher perspective that allows the researcher to understand and compare housing systems in different countries. The higher perspective does not change the sub-systems of housing provision but it does change the way in which they are deployed. So, here we have a series of contexts which have a relationship of higher to lower: SHP to the four sub-systems; and the systems context to SHP.

Fourth, different definitions emerge from different perspectives: the dwelling itself, housing provision, design and the meaning of home. It raises a question as to how these different definitions relate to one another, a question discussed later in this chapter.

Fifth, the definition these researchers arrive at is itself an expression of meaning, usually in words. It is an expression of an understanding or interpretation of other expressions of meaning. It links together expressions of meaning whatever form they may take – external events (states or actions) and internal events (states, actions, experiences etc.) and transforms them into another form of expression. So, Hulse and Burke define, in words, a social housing allocations systems with its three layers of activities: a strategic planning process, a primary rationing process and a secondary rationing process.

Sixth, when defining, these researchers are selecting elements that they regard as relevant to what they are investigating. So, for instance, the ABS Census Dictionary (2011b) defines a dwelling as follows: 'In general terms, a dwelling is a structure which is intended to have people live in it, and which is habitable on Census Night. Some examples of dwellings are houses, motels, flats, caravans, prisons, tents, humpies and houseboats.'

This definition selects particular characteristics: it is a 'structure'; its purpose is for 'people [to] live in'; and it is 'habitable'. The definition ignores its colour, its size, its design, the number of rooms, the different types of rooms and the materials it is made of. Similarly, when Hulse and Burke (2005) develop their analytical framework for social housing allocations, they select those elements that are relevant to allocations and ignore others, such as the life-cycle of the dwelling or the interests and hobbies of the applicant.

Seventh, the definitions proposed by these researchers are used in various ways. At the outset of a paper, researchers will define their terms and use these definitions as a way of controlling the research: to focus the research, to limit its scope and to specify what is relevant and what is not. So, a definition in a data dictionary will precede the collection of data and will determine what is collected. An outline of an analytical framework will precede the analysis of what is happening in a particular location or country. By defining the framework of analysis, it becomes possible to understand the characteristics of these elements and to compare them with other places and times. As illustrated by Ambrose (1991) and Lundqvist (1992), it also provides a framework for understanding and proposing future directions. In this way, a definition of housing has a heuristic role.

Eighth, these researchers have different approaches to defining housing.

1   A data dictionary definition is an interpretation of how a particular word or term is commonly used, and this common usage determines the range of its relevance and its limitations.
2   Some definitions describe something by using similar but more familiar terms. In the definition of dwelling above, the ABS define the term by referring to examples such as houses, motels, flats, caravans, prisons, tenants, humpies and houseboats.
3   Some researchers such as Mallett (2004), Dowling and Mee (2007), and Blunt and Dowling (2006) maintain a multidimensional definition of
4   'home' by referring to the understanding of home by many and varied participants.
5   Some definitions describe a sequence of events. For example, McNelis and Burke (2004) initially describe public housing rent-setting and various forms of rent-setting in community housing in Australia in terms of the processes used by SHOs to determine a rent. Later they generalise these descriptions into different types of rent-setting systems but stop short of developing a definition of rent-setting that encompasses all forms of housing in different countries.
6   Conceptual or analytical frameworks define housing (or some aspect of it) by working out those elements that are common to housing as it operates in different locations. The source of these descriptions may be some mixture of personal experience, interviews, documents and accumulated knowledge, with each participant making a contribution which depends upon their involvement in only one or a limited number of the elements of the framework. The researcher develops the framework by generalising the common elements in a group of instances.

Towards a more precise understanding of Interpretation

The shift to theory remains the central challenge for housing researchers. What such a shift entails is unclear to housing researchers. This section takes up the challenge posed by King, 'a questioning of what we mean by theory; what its possibilities and limits are' (2009: 51). At its core, the shift to theory requires housing researchers to distinguish between different types of questions, in particular between two related but different questions: 'what is housing?' and 'what role does housing play, i.e. what is its purpose?' This section seeks to make these distinctions and work through their implications for housing research. It is a long, complex and difficult section but one which is central to grasping not just a more precise understanding of Interpretation but also of distinguishing Interpretation from other functional specialties.[1] Without a more precise understanding of theory, the remaining chapters will make little sense. I begin with a discussion of the problem faced by the three types of definition – defining terms, conceptual frameworks and participant descriptions – used in housing research.

The problem of definition

The key question facing a scientific approach to housing is whether one or all of these three types of definition are adequate and, if not, whether there is an alternative approach to defining an object of investigation.

In social science, the problem of definition is often framed in terms of whether researchers should uniquely privilege the descriptions of participants or social agents (a constructivist, ethnomethodologist or grounded theorist viewpoint) or ignore accounts of behaviour given by the agents themselves and present an objective definition (critical realist viewpoint) (Meynell 1975: 65). In my view, a scientific approach to housing requires an objective definition, yet we cannot overlook or dismiss the descriptions of participants or agents. It is through their activities that housing is constructed. However, the issue is what in these descriptive definitions is relevant and what is not relevant to a definition of housing. And it is here that we can begin to distinguish between definitions that pertain to the everyday taken-for-granted world of common-sense living, on the one hand, and definitions that pertain to the world of science.

In Chapter 3, I noted how, in our everyday taken-for-granted living, we are oriented towards immediate practicalities and do not clearly distinguish between the types of questions we ask; we are concerned with things insofar as they contribute (or not) to our living, to our concerns and interests. As a result we are interested in understanding things insofar as we need to use them and insofar as they affect us or others. Such understanding will vary according to our experience and our perspective, our social and cultural background etc. Above I quoted Mallett (2004: 84) as follows:

> the term home functions as a repository for complex, inter-related and at times contradictory socio-cultural ideas about people's relationship with one

another, especially family, and with places, spaces, and things. It can be a
dwelling place or a lived space of interaction between people, places, things;
or perhaps both.

She recognises that when we define home or express the significance of home we
do so in terms of its value to us, its purpose in relation to us. What home is
becomes inextricably linked with its role in our personal history (and that of our
cultural group) and with habitual associations that have developed over a lifetime.
The strength of this association makes it very difficult to distinguish the role or
purpose of home from what home is, i.e. what it is that is serving that role or
purpose. Indeed, because home does have a role in our lives, we assume we know
what home is.

The data, such as that provided by Mallett, expresses this ambivalence between
what home is and its associated role or purpose in our living. Any attempt at
defining housing must take into account this ambivalence of many different
human expressions of its meaning or significance. If we are to develop a definition
of housing within a scientific context, our challenge is to sort through these
varying perceptions and understandings of housing and work out a definition
which critically distinguishes between what is relevant to what housing is and
what its role or purpose is.

As a way of taking this discussion forward, I will reflect upon one of the more
prominent theories of housing provision, the structures of housing provision thesis
as proposed by Ball and Harloe (Ball 1983, 1986, 1998; Ball et al. 1988; Ball &
Harloe 1992). SHP offers a starting point for a more adequate and complete theory
of housing, despite being the subject of extensive critical comment (Hayward
1986; Kemeny 1987; Oxley 1991; Lawson 2006; Dodson 2007). In reflecting
upon SHP, I am not seeking to develop an adequate and complete theory of
housing. Rather, I will propose three revisions to the structures of housing
provision that will provide a starting point for the development of such a theory:
a shift from methodological procedure to a theory; a shift from focusing on the
inter-relations between social agents to grasping the functional relations that
constitute housing; and a shift in general references to context to precise, defined
and related contexts.

Theory as an answer to a what-is-it-question

One of the difficulties that Ball and Harloe face in their defence of SHP is to
specify what it is. In the face of their critics, they opt to refer to SHP as 'research
methodology' or 'methodological procedure'. In doing so, they are seeking some
middle ground between the charge that SHP is empiricist and that it is a general
theory of housing.

While the SHP in each country is empirically derived, Ball and Harloe (1992) go
on to argue that the procedure is not empiricist because 'it explicitly recognises the
interaction between observation, theory and individual judgement' (p.5). But they
explicitly reject the view that SHP is a theory of housing; rather, it is 'a metatheoretical

concept or analytical framework which, together with other theories, may be of use in the examination of particular aspects of housing development' (p.3). Later they explain their reluctance to view SHP as a theory of housing.

> In that sense [as a methodological procedure] it is a theory but it is not a general theory of housing whose postulates can explain most housing-related problems. Moreover it is wrong to think that such a general theory can exist. 'Housing' involves many diverse social processes whose explanation cannot be encompassed within one grand theory...(p.6)

Their difficulty derives from their understanding of theory. For them, 'a theory of housing...produces from postulates a set of results claimed to have universal empirical generality' (p.4). In other words, from a theory we should be able to deduce what is empirically observable not only in one country but in all countries. It is in this way that the theory is verified.

This discussion reflects a tension or ambivalence within SHP – a tension between an inductive methodology with an empiricist priority for what can be observed, and a deductive methodology with an (idealist) priority for the preceding analytical or conceptual framework that a researcher brings. In my view, Ball and Harloe's understanding of theory is informed by the (mainly British) tradition of induction and deduction as procedures in science. The adequacy of this tradition is a question for the fourth functional specialty Dialectic. A phenomenology of research reveals that theoretical understanding is not just a logical process of induction or deduction. It also includes non-logical processes of inquiry, observation, discovery through insight and affirmation.

> [M]odern science derives its distinctive character from this grouping together of logical and non-logical operations. The logical tend to consolidate what has been achieved. The non-logical keep all achievement open to further advance. The conjunction of the two results in an open, ongoing, progressive and cumulative process.
>
> (Lonergan [1972] 1990: 6. See also Meynell 1998).

What is clear for Ball and Harloe, however, is that SHP is a 'methodological procedure':

> SHP...is meant to provide a productive framework within which housing-related issues can be examined. It enables empirical material to be ordered and inter-related in a particular way and by doing so closes off certain lines of explanation. The form of the ordering of empirical material enables consideration of the dynamic context of housing-related issues as it gives priority to the inter-relations between the institutions involved in a SHP. The closure aspect is important as it provides a basis for criticising a number of widely prevalent approaches including consumption-orientated and/or liberal-interventionist explanations of housing problems...(p.6)

In many ways this 'methodological procedure' resembles the role of theory as a heuristic, an anticipation of the structure of an answer to a question: SHP anticipates that the inter-relations between the social agents involved in housing provision (in production, exchange and consumption) is decisive for the form of housing provision in a particular country.

This raises two questions: first, on what basis is SHP proposed as the *relevant* conceptual or analytical framework for housing provision? Second, is SHP fixed or can it develop as a framework? I would contend that SHP can only function as a methodological procedure if it is some answer to a what-is-it-question. Further, that the debates about the priority of consumption or production and about provision have a common context – housing – and they can only be resolved by recognising that the debate is about answering a what-is-it-question.

Ball seems to indicate this when in an earlier article introducing SHP as an alternative theorising of housing provision, he asks: 'To put the matter as a naive question: what is studied in the study of housing?' (Ball 1986). Could not this question be framed more simply as: what is housing? An answer to this question is what is studied in the study of housing. Ball and Harloe, for reasons that require further exploration, seem reluctant to identify SHP as an answer to a what-is-it-question. Some of this reluctance may stem from collapsing four questions – the empirical question, the theoretical question, the historical question and the evaluative/critical question – into one rather than distinguishing them.

The term 'theory' may be used in any number of ways. It can, however, be given a more precise meaning if it is understood as an answer to a what-is-it-question. Now it could be objected, on epistemological grounds, that we cannot answer a what-is-it-question. All I am doing here is 'extending' the approach of researchers such as Mallett who note the differing views of participants on the meaning of home and seek to maintain a multidimensional view. In doing so, they are also answering a what-is-it-question: what does 'home' mean for this or that participant? I am proceeding by inquiring into this same data to reach for an understanding of 'home' that is presupposed in data such that Mallett can report on these views under the umbrella of 'home'. If we can understand the meaning of home for each participant, then it must also be possible to grasp what 'home' is.

By answering this what-is-it-question a researcher develops some control of the process of research. By defining more precisely what is being researched we can go on to investigate its occurrences (and whether the definition is adequate), understand how this 'what' changes over time, how it is used, critique its limitations and propose how we can improve it.

Theory as functional relations

Ball describes a structure of housing provision as 'an historically given process of providing and reproducing the physical entity, housing; focusing on the social agents essential to that process and the relations between them' (1986: 158). These social agents are the institutions involved in the production, exchange and consumption of housing. It is the inter-relations between them that create and sustain the particular structures of housing provision in each country (Ball & Harloe 1992: 3). Within actual structures of housing provision 'there are likely to be varying relations of power and of domination and subordination' (Ball 1986: 158).

In our everyday taken-for-granted living, anticipating the actions, responses, attitudes and motivations of social agents (whether individuals or institutions) are particularly important as we work out how to negotiate our world, how to get what we need or want to achieve, how we develop solutions to immediate practical concerns. An understanding of the 'varying relations of power and of domination and subordination' among the social agents involved in housing provision helps us work out with whom to align to bring about some change in policy direction and/or some benefit to ourselves and the groups to which we belong. This understanding, however, is not an understanding of housing *per se* but rather an understanding of the social agents involved in housing provision and how they are using housing (and exercising their power) to achieve something else that is in their own interests.[2]

As a prelude to a discussion of the question 'what is housing?', I would recall the discussion on a scientific approach to housing in Chapter 3. Within a scientific context, our orientation is towards asking and answering questions. The researcher not only prolongs the questioning process with more searching and more critical questions. It is a context within which we distinguish between different types of questions, each of which anticipates a different type of answer.

So, what is housing? What type of answer does this type of question anticipate? By what method do we answer what-is-it-questions? We can begin by noting some curious features about asking this question. First, our question 'what is housing?' names what to us at the outset is something that, in one sense, is already known (for how can we ask a question about something that is completely unknown) but, in another sense, is an unknown that we seek to understand (that is why we are asking a question). Confusingly, such naming is the beginning of our research, not the end. Just because we can name something as 'housing', regularly use the word 'housing' or even describe 'housing', it does not mean that we understand what housing is. As Melchin (1999: 62) notes, this naming does serve a heuristic function.

> The investigator can name the object. But initially the name has no meaning, no familiarity, no intelligibility. The function of the name is heuristic. The name does not serve to classify the object but only to point to it as an object that can be experienced in some way or another but remains to be understood. 'Let the object be named *a*,' where *a* can be any set of marks, squiggles, letters, or characters as long as it is not presupposed that we know what *a* 'means.'

By naming what we are inquiring about, we are pointing to something beyond our current understanding. The naming, in some sense, enables us to hold ''housing' while we inquire into it.[3] Melchin goes on:

The next step is a little more complicated. The investigator must turn his or her attention to the empirical occurrences of *a* and to whatever experiential evidence can be gathered about *a*...

In turning to occurrences of what we name 'housing', we presuppose some understanding of housing. It may be one of the types of definition outlined above (a definition of terms, a conceptual framework that identifies and links particular aspects of housing and/or various participant descriptions (including our own)) or, as outlined below, a current explanatory definition or theory of housing. Our current understanding of housing serves as a heuristic, it guides our research. We seek some better understanding because we are puzzled by something, something doesn't make sense, our current definition is incomplete.

To reach some better understanding, we turn to what we currently understand as occurrences of 'housing', we turn to and gather data on these occurrences of housing. We come with a series of questions that are guided by our originating what-is-it-question: what is associated with housing?, what other things, events, processes could be associated with housing? We seek:

clues to the appropriate context or perspective in which *a* is to be understood. Is *a* an operation or the result of an operation? Is it a unity or a manifold? Does it have a structure? Where does it begin and where does it end?...Is *a* to be understood in relation to *b* or in relation to *c*?...By shifting context and perspective, trying to bring one or another set of questions to bear on *a*, listing the data, juggling it around, rejecting one perspective in favour of another, performing endless operations in controlled settings to test possible sets of questions and answers, the investigator moves more or less slowly towards a discovery.

(Melchin 1999: 62–63)[4]

In reaching for a discovery, we explore and mess with the data; we use 'fantasy and lateral thinking' to make sense of this data; we distinguish events and activities; we highlight some as relevant and others as not; we notice that some are associated with each other in time, in place or in both time and place; we distinguish sets of events and activities (elements), we categorise them into different sets and then recategorise the events and activities into different sets; we make suppositions and test them; we propose hypotheses about the relevant sets of activities and their relationships, rejecting some as inadequate, considering new possibilities etc. All the while we are stretching ourselves, reaching for some insight into the data, reaching for a discovery that will answer our question and make sense of the data. We are reaching for one of those 'aha' moments, a 'eureka' moment when suddenly all the pieces come together, when all the significant and relevant data to answer our question is included and all the misleading and irrelevant data is excluded and when we understand just how certain sets of activities relate to one another (just as the detective suddenly puts all the relevant significant clues together, excluding those which have been misleading and irrelevant).[5]

A what-is-it-question anticipates that some intelligibility will emerge from the manifold of events and activities in our experience. The researcher is looking beyond associations between housing and other things. In the face of a manifold of diverse data, the researcher uses 'fantasy and lateral thinking' to reach for a discovery, an insight which 'grasps a relation or set of relations that define *a* in terms of its appropriate context of other elements' (p.63). Through insight an understanding of the significance of the other elements in relation to *a* is reached.

Here I am making an important distinction between an association between data and a correlation. An association occurs when events are externally related through their remote or immediate proximity in time and space. A set of events or properties is associated when they regularly and conjointly occur in space and time. A correlation, on the other hand, is an internal or systemic relationship between events that is grasped through insight. A set of events or properties is correlated where the occurrence of an event or property is conditional upon the occurrence of two or more of these events. An insight or set of insights grasps the systematic relationship between a set of events that condition the occurrence of another event.

Confusingly some researchers use these terms interchangeably. For example, Blaikie notes: 'Two variables are said to be associated if the values of one variable vary or change together with the values of the other variable; the variables are said to be co-related' (2003: 89f, 306). Indeed, some researchers may hypothesise or theorise about associations (for example, Dockery et al. 2008: 14f). But just because there is an association in time and place does not mean that the relationship between events is a relationship of significance. Associations in time and place can play a role in discovering the significance or meaning of these events. Moreover, an association in time and place is a necessary condition for this significance or meaning. However, this significance or meaning is not equivalent to discovering the association between the occurrence of these events in time and space. In other words, an association in time and space is not the only condition for discovering the significance or meaning of these events.

Understanding or insight is that creative moment of 'fantasy and lateral thinking' when I grasp a relation or set of relations (a correlation) between the essential, relevant and significant elements such that I grasp a possible answer to my question: what is housing? The relation or set of relations is not simply one of time/place association but rather a correlation where the regular occurrence of housing is systematically related to the regular occurrence of other events in a certain pattern. The insight unifies these events/activities/processes/elements – it implicitly defines their meaning in their relationship to one another; it is a higher viewpoint on these events/activities/processes/elements, a viewpoint which integrates them into a larger whole.

But we are not just interested in what might be, in possibilities. There emerges, then, a new question, an is-it-question: is this housing? Does this formulated hypothesis account for all essential, relevant and significant data, and only the data that is essential, relevant and significant? Is it actually so? The hypothesis has its conditions: whether enough attention has been paid to the data; whether the data is sufficient; whether all the relevant questions concerning the data have been asked. This is-it-question weighs up whether these conditions have or haven't been met and makes a judgement about the hypothesis' certainty – it is so, it is likely to be so, it may be so, it isn't so. Is-it-questions move us beyond just any understanding of the data to our best understanding. The criteria by which we make our judgement are whether we are satisfied that we have met the conditions for correctly answering the what-is-it-question. Insofar as we have not met these conditions, we raise new questions (we may raise them or others may raise them) which seek new data and new insights and push us to new hypotheses.

So an insight that answers the question 'what is housing?' will be a viewpoint that grasps a complete set of related activities (elements) whose occurrence are the conditions for the occurrence of housing. Thus, we are seeking an insight into the elements that bring about or constitute housing. This insight or discovery systematically relates the occurrence of elements with the occurrence of housing. These elements are the conditions for the occurrence of housing (Melchin 1991, 1994, 1998, 1999, 2003).

Let me return now to the structures of housing provision thesis proposed by Ball and Harloe. What is relevant or significant to a theory of housing provision is the set of related elements which constitute housing provision. In doing so, the elements are related in that they play a role or purpose or function. The motivations, attitudes and interests of social agents are not relevant or significant. What is relevant and significant for constituting housing provision is that certain activities/ processes/elements occur and that they occur in a certain relationship to one another. This set of elements is internally related to one another.

A theory of housing will have certain characteristics. First, it is an explanatory definition, a complete set of related elements that constitute housing, that explain the occurrence of housing. Other types of definitions – descriptive, analytical frameworks and participant – select certain elements of housing that are relevant to a discussion. A theory of housing heads towards a *complete* set of elements that are functionally related and in so doing distinguishes housing from other things.

Second, a theory of housing is abstract, where abstract has a particular meaning. From the vast and diverse range of manifold events and activities that can be associated in time and place with housing, a theory of housing selects that set of related elements which is relevant and significant to the occurrence of housing. It abstracts from or overlooks many events and activities which are neither important nor relevant nor significant to actually constituting housing.[6] Insight creatively leaps from associations in time and place between data to a grasp of a systematic correlation between the occurrence of a set of related elements and the occurrence of housing.[7]

What is being proposed here is not something ideal or utopian but rather the actually occurring processes that constitute the occurrence of housing. Insight moves beyond *associations* in time and place to *correlations* that grasp the complete set of related elements that constitute housing. By selecting what is relevant and significant, insight 'enhances' the data.

Third, a theory of housing is universal, invariant and normative. Unlike analytic and conceptual frameworks, this theory is not distilling what is common among various instances of housing. Many things may be commonly associated with housing but they may not be essential, relevant and significant. A theory of housing heads towards that complete set of related elements that constitute all occurrences of housing at whatever time and place. In asking 'what is housing', we are seeking that set of related elements that are relevant and significant to constituting housing and that distinguish housing from everything else regardless of its time and place. While our understanding of housing is ever incomplete and affirmed as more or less probable, while the expression of housing may vary from person to person and from culture to culture, the goal is always to grasp the set of related elements that constitute housing in whatever time or place it occurs. In this way, a theory of housing is universal, invariant and normative.

Fourth, as universal, invariant and normative, a theory of housing can function as a heuristic. It can guide the functional specialty Research as it turns to and gathers data. It can guide the functional specialty History as it seeks to grasp the dynamic of change in actual operating housing systems. It can guide the functional specialty Dialectic as it understands the ways in which individuals and groups promote their own interests and aggrandise themselves. It can guide the implementation functional specialties: Foundations, Policies, Systematics and Communications.

Theory within related contexts

Ball and Harloe (1992: 5) 'recognise the context-bound nature of housing provision' and that 'differences between SHPs are influenced by historical, geographical and physical features'. Moreover, 'each social science has its own theories to explain particular social processes' (Ball 1986: 150). In this context, SHP 'can exist only with other types of theory as part of the set of analytical tools used to investigate empirical questions. The set would also include amongst other things theories of social and economic processes and theories of data analysis and evaluation' (Ball & Harloe 1992: 13).

As noted previously in this chapter, Burke superimposes a 'systems context' on SHP in order to understand the particular structures of housing provision in Australia compared with other countries. As Burke and Hulse (2010: 825) note, these are

> embedded in a broader institutional context made up of the particular economic, social, demographic, administrative, legal, environmental and political processes which define Australia and shape the form, timing and direction of housing policy and problems. This institutional context also embodies the dominant norms and values of society...

An understanding of the context is important for a more complete understanding of housing provision, where 'complete' means an understanding not just of housing provision as 'a general theory' but also an understanding of SHP as it operates in each country. Context, then, is a way of understanding the particularities of SHP. The current understanding of context within SHP, however, suffers from two weaknesses. First, it is largely based on associations between a particular type of context and a particular type of SHP. For example, Joris Hoekstra (2010) and Kath Hulse (2003) have both sought to relate housing and politics by applying Esping-Andersen's (1990) typology of welfare state regimes to housing. Second, the current understanding relates SHP to context in a piecemeal way. It does not seek a more comprehensive understanding of context, one which incorporates (i) the various participant definitions of housing, such as the multi-dimensional view of home outlined in the work of Mallett (referred to above) and (ii) the various disciplinary approaches. By relating and incorporating these perspectives into a theory of housing, we would have a more adequate and complete theory, one which will provide a more adequate and complete heuristic for understanding an actual operating housing system with all its particularities. A more adequate and complete heuristic of housing becomes critical when seeking answers to other types of questions: the empirical question, the historical question, the evaluative/critical question, the transformative/visionary question, the policy question, the strategic question and the practical question.

It is this question of relationship between contexts that is not well explored by Ball and Harloe. Further, as noted above, they are reluctant to call SHP a theory because they are unable to deduce from it what is empirically observable.

The context of housing

In answering the question 'what is housing?', I proposed a particular type of answer, an answer which is grasped through an insight which distinguishes housing from other things by discovering the systematic relations between selected empirical data. By selecting certain data as relevant and significant, insight 'passes over' other data as irrelevant and insignificant. I further noted that this answer is abstract. As a consequence, a theory of housing is universal, invariant and normative.

There is one further consequence. While the complete set of related elements that constitute housing is fixed, the particularity of the elements themselves is not. For example, if we consider the set of related elements that constitute a dwelling, it would include elements such as materials and design in certain relations. The particular materials that, in part, constitute a dwelling may, however, vary. They could be one or a combination of timber, stone, bricks, concrete, glass, plastic, steel, aluminium, ice, brush, thatch etc. The particular designs that, in part, constitute a dwelling may include any number of rooms in different configurations with differing purposes according to social, cultural and personal preferences (see for example, Pierre Bourdieu's essay on the Kayble house (1979)). The range of these materials or designs is not unlimited; rather, there is a range of possibilities such that the particularity of the dwelling may vary.[8]

A theory of housing, then, will admit a range of possibilities. In this way, the particularity of housing will vary. It is this particularity of the related elements that we seek to explain by appealing to the context within which housing is operating. Let me expand on this further.

We have proposed that a theory of housing is a complete set of related elements. The occurrence of housing is conditional upon the occurrence of this complete set. For Ball and Harloe, these elements are production, exchange and consumption.[9] The occurrence of each of these also has its conditions. So, we can ask about the conditions for the occurrence of each element and these conditions will be understood as answers to such questions as: What is production? What is exchange? What is consumption? In answering questions such as these, we turn to the data seeking to discover the systematic relations between selected empirical data (the relevant and significant data) that will constitute production, exchange or consumption.

The relevant and significant data will not be the same data for each. The complete set of related elements that constitute production, exchange and consumption will be different – each set will distinguish one element from the other.[10]

There is a further point to note. Just as the answer to the question 'what is housing?' is abstract, universal, invariant and normative, the answers to the questions 'what is production?', 'what is exchange?' and 'what is consumption?' are answers that are abstract, universal, invariant and normative. They pertain to any instance of production, exchange and consumption at any time or place. Together they play a role or function in constituting housing. However, housing is one particular good which is produced, exchanged or consumed, and to constitute housing these elements are particularised – adapted and moulded – in such a way that they constitute housing and not something else, they are particularised in view of the demands of housing (rather than any other good or any other mode of property development whether commercial, retail, wholesale etc.). At the same time, because production, exchange and consumption themselves are conditional, the extent to which they can be particularised, adapted and moulded is limited. In this way they limit the range of possibilities for housing. Production, exchange and consumption are particularised in such a way that together they will achieve a particular purpose[11] – constituting housing. Production, exchange and consumption as constitutive of housing are related to housing as lower purpose to higher purpose.

Further, just as housing is a higher purpose in relation to the elements (production, exchange and consumption) that constitute it, so too housing itself can play a role in an even higher purpose – it can be one of a number of conditions or elements for the achievement of that higher purpose. So, for instance, by asking 'what is an economy?', 'what is a society?', 'what is a culture?', 'what is home?', we find that housing plays some sort of role either directly or indirectly in creating a standard of living, effective agreement between individuals and groups, a common way of life and a place to belong.

There is, then, an ordered set of relationships, a type of hierarchy of relations between: the sets of related elements that are the conditions for the occurrence of production, exchange and consumption; the set of related elements (production, exchange and consumption) that are the conditions for housing; the set of related elements (which include housing) that are the conditions for a standard of living; the set of related elements (which include a standard of living, of which housing is a condition) that are the conditions for effective agreement between individuals and groups; the set of related elements (which include effective agreement between individuals and groups) that are the conditions for a common way of life; and the set of related elements (which include a common way of life) that are the conditions for a place to belong, a home, etc.

As the particularity of each of the elements within this ordered set of relationships can vary within a range of possibilities, it would seem that we could develop a series of related theories which form an ordered set of theories within a particular type of hierarchy where:

i   a level is a purpose which systematises lower level elements (and higher and lower reference this conditional relationship);

ii   the higher level and the lower level are related but not related systematically, because:

iii the higher level depends upon the lower level, i.e. the occurrence of a higher level purpose depends upon the occurrence of lower level purposes which themselves depend upon the occurrence of even lower level purposes. However, the occurrence of these lower level purposes does not depend upon the occurrence of higher level purposes;

iv   the higher levels select or particularise the lower levels in order to achieve their particular purpose, i.e. the lower level has a role or purpose in the occurrence of a higher level purpose but this role is particularised from a range of possibilities.

This ordered set of theories is a particular type of hierarchy, one not based on dominance but on mutual interaction.[12]

In speaking of context, I am distinguishing between two types of questions: what-is-it-questions and purpose-questions. The first type is concerned with what housing is, the second with how housing is used for some purpose or other. The social agents involved in the production, exchange and consumption of housing are motivated by some purpose or other, and use housing to achieve it. This purpose is not housing but rather something else in which housing plays some role or purpose. The purpose-question, then, is not a different type of question. Rather, it is a different what-is-it-question. The 'what' here is one in which housing is one of the set of related elements.

One of the difficulties in developing a theory of housing is that of distinguishing between what is essential, significant and relevant to answering the question 'what is housing?', and what is essential, significant and relevant to answering other what-is-it-questions in which housing plays a role, but not the only role, in constituting something else. The development of a theory of housing requires that some precise distinctions are made and that sets of related elements are precisely understood. A more developed theory will not only grasp the set of related elements that constitute housing but also grasp the way in which its particularities (or properties or characteristics) are determined by the role that housing plays in constituting a standard of living, an economy and a society etc.

This discussion of context is an attempt to understand the particularities of an actual operating housing system through a series of theories which play some role in constituting housing or in which housing plays some role. As Lonergan observed: 'modern science uses universals as tools in its unrelenting efforts to approximate to concrete process' ([1968] 1974: 104). As an example, he approximated the movement of the planets through a series of theories, as follows:

> Newton's planetary theory had a first approximation in the first law of motion: bodies move in a straight line with constant velocity unless some force intervenes. There was a second approximation when the addition of the law of gravity between the sun and the planet yielded an elliptical orbit for the planet. A third approximation was reached when the influence of the gravity of the planets on one another is taken into account to reveal the perturbed ellipses in which the planets actually move. The point to this model is, of course, that in the intellectual construction of reality it is not any of the earlier stages of the construction but only the final product that actually exists. Planets do not move in straight lines nor in properly elliptical orbits; but these conceptions are needed to arrive at the perturbed ellipses in which they actually do move.
>
> (Lonergan 1974: 271–272)

In summary. On an initial view, a theory of housing is the complete set of related elements that constitute housing at any time or place. As a complete set, all the essential, significant and relevant related elements are included and the non-essential, insignificant and irrelevant elements are excluded. On a more developed view, a theory of housing is the complete set of related *variable* elements. As variable, the related elements admit of ranges of possibilities. Finally, a complete theory of housing can be understood as a complete ordered set of sets of related variable elements where the particularity of the complete set of related variable elements that constitute housing are ordered by sets of related variable elements that constitute other 'whats' in which housing plays some role. This ordering ensures that these 'whats' are achieved.

The complete context of housing

If we are to understand the context within which an actual housing system operates, then we need to understand the various purposes of housing as a series of 'whats' within which housing functions as one of the set of related elements. We can group these 'whats' into different types. In doing this, I want to bring together the full range of participant views of housing, not only

consumers but also builders, estate agents, financiers, managers, land developers, governments etc.

First, there is a range of differing views of participants on housing. This is
illustrated by the work of Mallett who sought to maintain a multi-dimensional
view of home where the meaning of home depends upon the historical, cultural
and personal context of each person. Her focus is on consumers. Above I noted
how definitions of home within the everyday taken-for-granted world of common-
sense living do not distinguish between what home is and its role in achieving
some other purpose. Home has a significance or meaning because it has some role
in participants' lives. The question, then, is how do these participant descriptions
fit within this hierarchy of 'whats' or, how do they order the particularities of
housing in such a way that these 'whats' will be achieved?

Second, these participant descriptions from consumers of housing can be
incorporated into a broader range of participant descriptions that are 'picked up'
by various disciplines such as the material meanings of housing as developed, for
example, by Gabriel and Jacobs (2008); the economic meanings of housing as
developed, for example, by Gibb (2009) and Ball and Harloe's structures of
housing provision (Ball 1986; Ball & Harloe 1992; Ball 1998); the political
meanings of housing as developed, for example, by Kemeny (1992), Bengtsson
(2009a) and Dodson (2007); the cultural meanings of housing as dwelling and as
home as developed, for example, by King (2004, 2005, 2008) and Blunt and
Dowling (2006) respectively; and the personal meanings of housing (or home) as
pointed to, for example, by Clapham (2002, 2005), King (2004, 2005, 2008) and
Mallett (2004).[13]

At the beginning of this chapter, I included a quote from Kemeny in which he
argues that the social science disciplines (sociology, economics, psychology,
politics, geography and history) dimension out aspects of social structures. The
challenge, however, is to develop a theory of housing that will hold together these
diverse dimensions and views of housing, that will take account of the diversity of
participant descriptions as incorporated into different disciplines and that will in
some comprehensive and integrated way relate this diversity of viewpoints. Such
a theory would be a more adequate theory of housing (but also more complex) and
would provide a more adequate heuristic for the further study of housing in later
functional specialties.

Here it is beyond my scope to develop such a theory. My primary interest is a
more precise understanding of the functional specialty Interpretation. So I simply
want to point to a way in which such diverse viewpoints could be integrated into
one theory which seeks to embrace 'housing' in a single view.[14]

The current approach of many social science disciplines to theory complicates
the question of interdisciplinarity because most social sciences, particularly
economics, are notable for their attempts to seek explanations by relating the
occurrence of events to human motivations and attitudes such as self-interest, an
approach rejected by the notion of theory outlined above. This notion of theory
differentiates different levels of a hierarchy of relationships according to their
purpose. A theory of housing is not just a complete set of related variable elements

but a complete ordered set of sets of related variable elements where different levels are differentiated and then related according to their purpose. Answers to a series of what-is-it-questions anticipate that a complete theory of housing will relate many different purposes.

The notion of theory proposed here suggests a more radical recasting of social sciences such as economics, politics, cultural studies, personal identity etc.[15] The difficult question is whether they can be differentiated and then related to one another according to their different purposes.[16]

A more adequate theory of housing would begin by recognising that economic, political, cultural and personal identity constructions are responses to: (i) our recurring need to produce goods and services out of which emerges a standard of living, (ii) our recurring need to reach effective agreement on how we co-operatively work together to produce this standard of living, (iii) our recurring need to discover, express and develop meaning and value in our living and (iv) our ever-present desire to transcend our current identity and to live in a larger and better world.[17]

The first recurring need is met through an economy. What is an economy? An economy is 'the totality of activities bridging the gap between the potentialities of nature, whether physical, chemical, vegetable, animal, or human nature, and, on the other hand, the actuality of a standard of living' (Lonergan [1942–1944] 1998: 232). One of the constituent elements of an economy is technology, the processes whereby society develops the know-how to transform the potentialities of nature into goods and services.[18] The second recurring need is met through a polity. What is a polity? A polity is the processes whereby, in an ever-changing environment, a society concretely reaches effective agreements as to who does what with what benefits or costs (such as how income and wealth are distributed). It encompasses decisions reached by individuals, by organisations, by businesses and by governments. These decisions are institutionalised in the taken-for-granted ways of doing things between people, in the policies and practices of organisations and businesses and in the laws and regulations of governments.[19] The third recurring need is met through a culture. What is culture? A culture is the inherited but ever-changing set of meanings and values that inform a common way of life. These are institutionalised in customs and mores, in arts and literature, in artefacts and materials, in aesthetics and architecture etc.[20] The fourth ever-present desire relates to personal identity. What is personal identity? It is the long process of developing an understanding of self, of attaining personal meanings and values, and of taking responsibility for who I am and who I become.[21]

These four are related in that they form a hierarchy in which the higher purpose depends upon lower purposes and the higher purpose systematises the lower purposes such that not only are the lower purposes achieved but the higher purpose is also achieved. So, technology evokes (that is, provides the conditions for) and is systematised by an economy, an economy evokes and is systematised by a polity, a polity evokes and is systematised by a culture, and a culture evokes and is systematised by persons who, in the measure that they have critically appropriated their tradition, can enlarge their own lives and those of others and become the source of new technological meanings, new economic meanings, new political meanings and new cultural meanings for future generations.

For example, a technology of housing (building materials, building techniques etc.) evokes and is systematised by an economy of housing (the processes that organise machines, equipment, labour, land, building materials, finance etc.). An economy of housing evokes and is systematised by a polity of housing (agreements about the division of labour, the division of finance and wealth, how people access and occupy dwellings). A polity of housing evokes and is systematised by a culture of housing (values such as equity, privacy). A culture of housing evokes and is systematised by persons who to express their identity through where they live, with whom they live, the style of their dwelling, the functionality of the dwellings, the artifacts they include in their dwelling.

Each discipline, whether technology, economics, politics, cultural studies or personal identity, has a particular focus and seeks to understand and explain a particular aspect of the whole. Each has a particular interest in the actual operating processes through which each particular society achieves an actual standard of living or actual agreements as to how things are done or actual meanings and values that inform a common way of life or actual personal identities. It is only by integrating these purposes that the whole can be grasped. What is actually operating and being achieved, no doubt, has its limitations, indeed, may be abusive of individuals. The integration of all these purposes explains what is being achieved. It is then that we can begin to assess the limitations of this achievement.

Figure 5.1 is a speculation on the set of related variable elements that might constitute different levels of understanding within an interdisciplinary approach to housing.[22] Vertically (as noted in the first or left-hand column), the figure distinguishes five disciplines: technology, economics, politics, cultural studies and personal identity.[23] The corresponding purpose of each discipline is outlined in the second column. A speculation on a possible set of related variable elements that constitutes this purpose is outlined in the third column. Each discipline contributes to a theory of housing and, thus, a more adequate heuristic.

As illustrated in the figure, housing as a component of a standard of living and as a product of an economy is embedded within the economy. For Ball and Harloe, its related variable elements are production, exchange and consumption. The task for an Interpreter is to work out the levels and their relations within the hierarchy as ordering or systematising the particularities of the set of related variable elements that constitute housing as one component of a standard of living: the range of possibilities in the set of related elements that constitute a technology sets certain limits for each of the elements of production, exchange and consumption; the achievement of housing plays a role in a standard of living and an economy orders or systematises the range of possibilities among the elements that constitute housing in such a way that a standard of living is achieved; the achievement of an economy plays a role in the functioning of politics (and within this economy, housing has its role), and politics orders or systematises

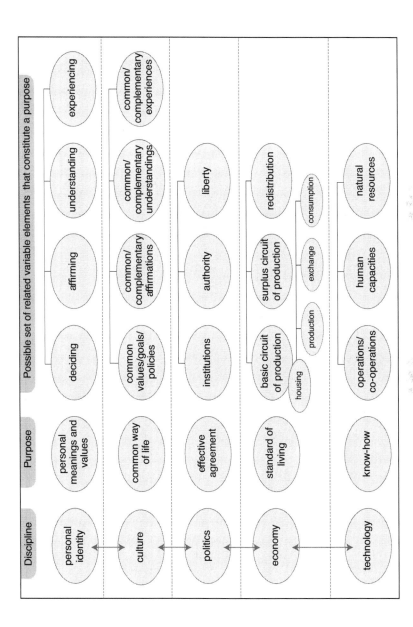

Figure 5.1 An interdisciplinary approach to housing: disciplines, their corresponding purpose and a speculation on a possible set of related variable elements that constitute their purpose

the range of possibilities among the elements that constitute an economy in such a way that individuals and groups reach agreement on who benefits and who loses;[24] the achievement of a politics plays a role in the functioning of a culture (and within the economy and politics, housing may play some role), and a culture orders or systematises the range of possibilities among the elements that constitute a politics in such a way that a common way of life is maintained; the achievement of a common way of life plays a role in the development of a personal identity (and within the economy, politics and culture, housing may play some role), and each person orders or systematises the range of possibilities among the elements that constitute a culture in such a way that the life of each person is meaningful.

What is being proposed here is a larger and more adequate heuristic for understanding housing at particular times and places. As a heuristic, this theory allows us to understand any particular actual housing situation or actual operating housing system. The goal of an adequate theory of housing is to develop a complete ordered set of sets of related variable elements that will constitute housing and its particularities. An actual operating housing system will be an ordered set of sets of related actual elements, i.e. the elements and their relationships as they actually occur.

By recasting various disciplines according to their purpose, it becomes possible to integrate them within a larger framework and in this way we can develop a much more adequate theory of housing. It is towards this goal of a complete integrated framework that incorporates and relates all aspects of housing that the functional specialty Interpretation is oriented.

Illustration: towards an indicative theory of social housing rent

The following is an indicative illustration of a theory of social housing rent. It is indicative because it still requires much work. Despite this, it does illustrate the notion of theory outlined above. It highlights the structure of theory as a complete ordered set of sets of related variable elements. It shows how the differing purposes of rent can be systematised into a set of theories within a particular type of hierarchy that can explain actual operating rental systems.

Social housing rent serves a number of purposes. We advert to these purposes when we talk about increasing rents to maintain the financial viability of an organisation; setting rents which are affordable; comparing rents on dwellings so that they are equitable; increasing or reducing rents on dwellings to encourage tenants to make different decisions about the dwellings they want; or comparing the housing costs of social housing tenants to owner-occupiers or private renters so these costs are equitable.

Now each of these purposes is achieved by some method, some series of activities and processes – a yet to be determined set of related elements. The initial question is: what is social housing rent? This can be followed by series of further questions regarding the role of social housing rent within the achievement of other purposes. Each of these other purposes is constituted by a set of related

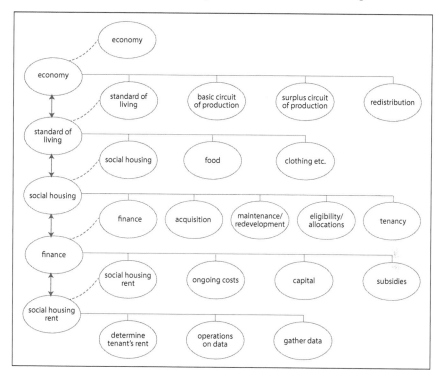

Figure 5.2 A theory of social housing rent

elements in which rent is directly or indirectly one of the elements that constitute this purpose.

Within a theory of social housing rent, each purpose forms a context and the ordered set of theories is a series of related contexts. The following outlines a theory of social housing rent within five contexts: social housing rent, finance, social housing, standard of living and economy. These are outlined schematically in Figure 5.2 (and should be read from the bottom upwards) with the lower context as one element in a higher context.

A first context: what is rent?

Our initial question is: What is rent? Or, what is the set of related elements that constitute a rent?

Rent is constituted by a set of three related elements (each with their own specific purpose) whose endpoint is a rent for each tenancy:

1   Gathering specified sets of data. Examples of datasets include: the prices at which dwellings with determined characteristics are sold and/or rented; the costs of acquiring and managing dwellings; types of tenant incomes and their

purposes; and tenant characteristics (for example, income and household type).

2   Specific operations are carried out upon these specified sets of data resulting in a set of criteria for determining rent for each tenant. These operations include collating, averaging, weighing, comparing etc. according to specific characteristics, and

3   Application of the criteria to a specific tenancy.

These three processes are serially related and dependent upon one another with each achieving something that is essential to social housing rent. Together they determine a tenant's rent. But what is also important to note is the range of possibilities within each of the stages: a range of possible sets of data; a range of possible combinations of operations, a range of possible criteria for determining the rent of each tenant. The range of possibilities points to the flexibility in determining a specific rental system according to the environment within which it operates. As a result, we have a variety of possible rental systems including: property rental systems (cost-rent, market rent, market-related rent) and household rental systems (income-related, subsidy-related and flat) (see McNelis 2006, 2009b).

This theory of rent outlines a structure which relates three key elements such that together they constitute a rent. Within that structure are a range of possibilities.

A second context: the role of rent within the social housing finance system

The rental system determines an individual rent but it can do so in a range of possible ways with a range of results. So on what basis are the possibilities arranged in a particular way such that a particular result (in addition to the determination of a tenant's rent) is achieved?

To ask this question is to shift to a different context or to a higher viewpoint. From this viewpoint, the rental system achieves not just a particular purpose – the determination of a tenant's rent – but contributes to another higher purpose. The question is about the role or function of the rental system within a larger context. We can begin by considering the immediate context of the social housing finance system.

In this context, what is important is not the individual rent but rather the aggregate of the rent that results from the particular operative rental system. In this new context, a tenant's rent is determined in such a way that aggregate rent is sufficient to cover the ongoing costs of social housing. The purpose of aggregate rent, as distinct from the purpose of the rental system, is to meet the ongoing costs of social housing.

But two further aspects of the overall social housing finance system complicate this relationship: the way in which social housing finances its acquisition of properties (capital finance) and the way in which subsidies are delivered into social housing. It is these four elements – rent, ongoing costs, capital and subsidies – and their relationships which together constitute the social housing finance system and bring about social housing financial viability (McNelis 2006).

The purpose of the social housing finance system (and the recurrent processes that constitute it) is to maintain the financial viability of social housing. But again, just as the rental system is open to a range of possibilities, so too the finance system is open to a range of possibilities: the level and type of ongoing costs can vary; the types of capital finance arrangements and the conditions under which they are provided (and their subsequent impact on ongoing costs) can vary; the level and type of subsidies provided to social housing can vary. These ranges of possibilities also point to the flexibility in determining a specific finance system according to the environment in which it operates. Again, we have a variety of possible social housing finance systems within and between countries (see McNelis 2006, 2009b).

A third context: the role of rent within social housing

The finance system operates as one system within the larger context of the systems that constitute social housing. Possible systems that might constitute social housing are: finance, production/acquisition, maintenance/redevelopment, management/eligibility/allocations and consumption/tenancy.[25] It is these systems (with sets of related elements) that together constitute social housing.

While the specific purpose of the social housing finance system is to ensure the financial viability of social housing, each of the other systems also has a specific purpose, and the rental system may or may not play some sort of role in each of these systems achieving their specific purpose. For example, the eligibility/allocations system may determine the income of tenants and thus their capacity to pay rent. Housing acquisition, tenancy and maintenance/redevelopment systems may determine costs which must be covered by rent payments.

Moreover, the totality of these systems that together constitute social housing is a new context or higher viewpoint, and the rental system may play some role in the constitution of this totality of systems (over and above its role within the systems that constitute social housing).

A fourth context: the role of social housing rent within a standard of living

Social housing is one form of housing among others that ensures that households are housed as one element that constitutes a standard of living. Its role or purpose is determined within this context and there are a range of possible roles which social housing can play. It can provide adequate and appropriate housing or provide affordable housing or provide housing to a range of income groups or only to those with high needs or provide particular types of housing etc.[26]

Within this context, the rental system may have a number of possible roles. It may be a mechanism for achieving affordability such as in Australia, Ireland, Canada, USA and New Zealand where household rental systems operate (see McNelis 2006, 2009b). By relating housing costs to income it may determine the relationship between housing and a range of other elements that constitute a standard of living.

A fifth context: the role of social housing rent within an economy

Social housing and its constituent systems operate within the broader context of systems external to social housing. For example, the acquisition of social housing operates with the larger systems of the building construction industry and the real estate industry; asset management operates within the larger systems of building and repair; eligibility and allocation operate within the larger systems of the distribution of housing; finance operates within the larger systems of capital finance, operating finance and subsidies. Thus the systems that constitute social housing are not isolated entities but rather cycle internally and externally (Melchin 2003).

These larger systems of acquisition, maintenance/redevelopment, eligibility/ allocations, capital finance and operating finance operate within a broader economy. An economy is constituted by a basic circuit of production, a surplus circuit of production and a redistribution circuit from which emerges a standard of living.[27]

Three comments on this indicative theory of rent

Three comments may clarify some aspects of this indicative theory of rent.

First, this is an ongoing attempt at developing an adequate theory of social housing rent. It is still incomplete and open to further development. Other possible contexts include: technological (the technology required to implement a rental system), political (how we make decisions about a rental system), cultural (reflections on the common meanings of rental systems) and personal (reflections on personal meanings of rental systems).

Second, the theory of rent shifts away from descriptions of the various activities of particular people and away from the motivations, interests and perspectives of the players towards sets of related activities or processes. It also differentiates the purposes of rent and the systems that constitute these purposes and, as such, shows how these purposes function in relationship to one another. Further, it illustrates how a theory systematises sets of relations and possibilities with higher purposes dependent upon the lower, but the higher making demands on the lower so that the higher purposes are also achieved. This is possible because there are a range of possibilities for rental systems, a range of possibilities for finance systems, a range of possibilities for social housing etc. As a result, the lower level rental system systematises those elements that constitute rent and explains how the rent of a specific tenant is determined. However, the rental system does not explain the level of the rent, why the rent changes, what form of rent the tenant is subject to and whether the rent is equitable or affordable. In relation to the rental system, these elements are not related systematically. If we are to explain these elements, we have to turn to other purposes within which the rental system is one among a number of elements. These other purposes can utilise the rental system to achieve their purposes because the elements of the rental system are not set. Rather, there is a range of possibilities. As a result, the rental system has a flexibility that allows it to achieve other purposes. If we are to understand the coincidental particularities of the rental system, then we need a higher viewpoint, a new context which systematises further aspects of the data. The demands of these higher viewpoints determine specific elements of the rental system. So, the finance system determines that the aggregation of individual rent payments will ensure the financial viability of social housing and that rents are regularly reviewed as the aggregate level within other elements adjusts (due to external changes); the role of social housing within a standard of living will determine the type of rent and the extent of this type, i.e. whether a tenant is subject to a property rent, a household rent or combination thereof; the economic system will determine the cost at which adequate and appropriate housing will be provided and thus the overall level of rents.

Third, through a series of approximations, the theory of rent, outlined above, can explain any particular rental system. As a first approximation, rent is constituted by three elements (datasets; combinations of operations on the datasets; criteria for determining the rent of each tenant) and their relationships. Each element admits of a range of possibilities and, as such, they do not of themselves explain a particular rent. As a second approximation, then, the range of possibilities within each of the elements is further limited by the role that rent plays in the financial viability of social housing. As a third approximation, the range of possibilities within each of the elements is further limited by the role that rent plays in relation to the constitution of social housing (and the other elements that constitute social housing). As a fourth approximation, the range of possibilities is even further limited by the role that rent plays in relation to a standard of living (and the role it plays in providing affordable housing). And further approximations can be made and the range of possibilities further limited by considering the role that rent plays, if any, in relation to other possible purposes such as tenure neutrality, equity etc. In this way, the theory of rent will explain a particular rental system.

Implications for housing research

This chapter has sought to give a more precise meaning to the functional specialty Interpretation. It began by distinguishing, through a phenomenology of inquiry, between data and an understanding or interpretation of the meaning or significance of that data. By asking a what-is-it-question, a question of definition, we are seeking to understand some data, to grasp its meaning or significance.

Three genres of research – defining terms, conceptual frameworks and participant descriptions – are primarily concerned with defining housing (or some aspect of it). The starting point for definitions is ordinary social life. Definitions grasp the meaning of apparently disparate meaningful events; they operate in contexts which are related to one another; they are a new expression of other expressions of meaning; they are selective; they are used for different purposes – to control or focus our research, to limit its scope and to specify what is and is not relevant. Researchers have different approaches to defining housing: by defining how words are used with similar but more familiar terms; by working out those elements that are common to housing as it operates in different locations; and by participants defining the meaning of housing.

These descriptive definitions are inadequate. They are understandings that operate within the horizon of the everyday taken-for-granted world of common-sense living where housing is defined in terms of its value or usefulness to us. Descriptive definitions of housing 'run together' housing with other things that have become associated with it. They do not distinguish between what is relevant to a definition of housing and the purposes associated with it. A scientific approach, on the other hand, seeks to grasp the uniqueness of housing, how it is distinguished from everything else. In a scientific approach, a definition of housing is a particular type of answer to a what-is-it-question, one which anticipates the complete set of related variable elements that constitute housing. This answer is an explanatory definition of housing or, to use the traditional term, a theory of housing.

A theory of housing has *seven* notable characteristics. *First*, a theory of housing distinguishes housing from other things. *Second*, theory is abstract. It shifts from associations between events/activities/processes in time and place to correlations that grasp what is systematic, a complete set of related variable elements, regardless of time and place. *Third*, a theory of housing is universal, invariant and normative. It grasps that the complete set of related variable elements that constitute housing are the conditions for the actual occurrence of housing at any place or time. *Fourth*, while a theory of housing is universal, invariant and normative, the particularity of its elements can vary. Thus, a theory of housing consists of a set of variables which admits of a range of possibilities. It is this characteristic that gives it its *fifth* notable characteristic and allows it to play a role as a heuristic, as an anticipation of the structure of what will be understood at the completion of an inquiry. *Sixth*, a theory of housing distinguishes between what housing is and the role it plays as one of a complete set of related elements. These roles form a series of contexts as a particular type of hierarchy. Each level of this hierarchy is distinguished by a particular purpose. Each level systematises lower level elements yet depends upon the occurrence of these lower level elements. But each level also contributes to some higher purpose. This series of contexts allows for the full range of purposes of housing including various participant descriptions. *Finally*, a theory of housing presented here opens up the possibility of an interdisciplinary theory. This, however, requires a recasting of disciplines – technology, economics, politics, cultural studies, personal identity etc. – according to a particular purpose: economics, our recurrent need for goods and services to maintain a standard of living; politics, our recurrent need for effective agreement; cultural studies, our recurrent need for meaning and value in our living; and personal identity, our ever-present desire to transcend our current identity and to live in a larger and better world. These purposes form a hierarchy whereby a higher level purpose systematises the elements of a lower level purpose, and the elements of a lower level purpose contribute to the achievement of a higher level purpose. By relating these purposes to one another it becomes possible to develop a more adequate theory of housing, one which is more comprehensive, integrated and interdisciplinary. A theory of housing is adequate insofar as it accounts for all the possible roles that housing can play in the larger framework of living. It is this larger framework of living that provides the empirical data that needs to be understood.

Theory is not something remote from practice, from empirical evidence. Rather, it is the way in which we get to what is most significant and relevant. The development of theory is not some cold, rational process of deduction but rather a passionate commitment to discovery. It isn't an optional extra. It is a commitment to understanding that requires all our faculties of wonder, desire, puzzling, sensibility, feeling, attention to detail, imagination, insight, creativity, fantasy, formulation, deliberation and action.

The functional specialty Interpretation is *ecstatic* for the researcher moves from their original understanding of 'housing' to a new understanding of 'housing'. It is *selective* for out of all the available data the researcher selects only some of the data. It is *critical* for the researcher selects the data that is essential, significant and relevant to what is under consideration, excludes the data that is not essential, insignificant and irrelevant and may locate it within a different context. It is *constructive* for it integrates the data and provides a higher viewpoint on the data – what previously was only grasped as disparate and unrelated events, through a theory of housing is held together in one perspective. Our originating question 'what is housing?' presupposes some grasp of housing. At the very least it can be pointed to in some way either materially or symbolically or both. In this way our question is a *heuristic* which through further research and the discovery of an answer gives way to a new *heuristic* for future research (paraphrasing Lonergan ([1972] 1990: 188–189).

The introduction to this chapter outlined a dispute between those who propose that the role of housing research is to *apply* social theory to a housing issue and those who propose that its role is to *create* a housing theory. As a way of untangling these messy debates, the discussion in this chapter proposes that theory plays both roles. It has outlined the characteristics of a more adequate theory of housing as a complete ordered set of sets of related variable elements. This theory is *created* in the functional specialty Interpretation as an answer to the question 'what is housing?' It is a theory which anticipates the particularities of any actual operating housing system. All housing research begins with some understanding of housing. We can fail to be explicit about our understanding. We can use an inadequate theory. We cannot, however, begin our research with a blank slate or *tabula rasa*. The goal of the functional specialty Interpretation is to develop a more adequate and complete theory of housing. It is this theory that will provide the best way for further research. It is this theory that can be *applied* as a heuristic for grasping any actual operating housing system regardless of time and place. As a heuristic, it orients and guides the functional specialty Research in its search for events/processes/purposes which are not incorporated into the current theory of housing and in its aim to upset the current theory. As a heuristic, it orients and guides the functional specialty History as it seeks to answer the historical question; it guides and orients the functional specialty Dialectic as it seeks to answer the evaluative/critical question; and it guides and orients the implementation functional specialties as they seek to answer the transformative/visionary question, the policy question, the strategic question and the practical question.

The functional specialty Interpretation proposes a shift in the way housing is defined *from* descriptive definitions reached within an everyday common-sense framework *to* answering a theoretical question: what are the elements and their relationships that constitute housing? The answer to this theoretical question is an explanatory definition of housing, a complete ordered set of sets of related variable elements that constitutes housing. This explanatory definition distinguishes between those elements that are essential, significant and relevant to the constitution of housing from those which are unessential, insignificant and irrelevant to the constitution of housing. Housing is used for differing purposes by different groups. It is these roles or functions which are associated with housing and which order the particularity of the set of related variable elements that constitute housing as an actual operating housing system.

Notes

1   This more precise understanding of Interpretation is loosely based on (i) Chapters 1 to 10 of *Insight* (Lonergan [1957] 1992), in particular Chapter 8, Section 5 on emergent probability, (ii) Chapter 7 of *Method in Theology* (Lonergan [1972] 1990) and (iii) 'Cognitional Structure' (Lonergan 1967). These were supplemented by the writings of Kenneth Melchin (1991, 1994, 1998, 1999, in particular; 2003), James Sauer (1995) and William Mathews (1987).
2   While attitudes, feelings and motivations do not play in housing provision *per se*, I will show later in this chapter that they do have play a role in ordering the particularities of housing provision. Further, as discussed in Chapter 7 on the functional specialty Dialectic, the interests of social agents and the decisions they make about what is worthwhile can promote the development of housing or it can promote the self-aggrandisement of an individual or group.
3   See also Richard Feynman's distinction between being able to name something and knowing it (1969: 316ff).
4   In his discussion of empirical method, Melchin recognises the distinction between research and working towards a discovery. However, his text at this point does intertwine these two types of activities and I have removed those questions from this quote that relate to the gathering of data.
5   See how Peter Beer (2009) uses the Hitchcock film, *Dial M for Murder*, to illustrate Lonergan's cognitional theory.
6   These overlooked events/activities may, however, provide the conditions for the elements that constitute housing or they may provide the conditions for other things in which housing plays a role. In this sense, they are external to housing.
7   Two further examples may clarify this notion of abstraction. We cannot imagine a circle without imagining one with lines and of certain dimensions (maybe large, maybe small). Through insight we grasp a theory of circles that has neither lines nor is of a certain dimension. What is significant and relevant is a set of relationships between elements (centre and radii). This theory does not specify the length of the radii nor the location of the centre. These are irrelevant and insignificant. The theory grasps the relationship between the radii – they are equivalent – and the relationship of the radii to a centre.

A further example pertains to a dwelling. In some countries, the material used to construct a dwelling may only be timber or ice. This may have a very long history and so, in these countries, timber and ice are associated with dwellings. A theory of dwelling articulates the set of relationships between elements that constitute a dwelling. These elements include the materiality of a structure but abstract from the particular associations of timber and ice. We cannot imagine a house without such materiality whether it be timber, ice, brick, stone, concrete, steel or some mixture. But these particularities are not relevant or significant in a theory of a dwelling. Insight leaps to a certain type of materiality (one which can form a structure that is capable of providing shelter) as one of the elements of a dwelling that is relevant and significant.

8   In a similar way, a theory of a circle admits of a range of possibilities. By varying the length of the radii we still get circles, but of varying sizes.

9   For the purposes of illustration, I am assuming here that production, exchange and consumption are the significant and relevant elements that constitute housing. Whether they form an adequate theory of housing is a matter for further investigation.

10 Burke and Hulse (2010: 827ff), for instance, proposes that the relevant elements for production are land ownership, land development and house building; the relevant elements for exchange are the finance industry and the real estate industry; the relevant elements for consumption are the market for owner-occupied housing and private rental, non-market housing (social housing), type of housing stock and geographical concentration. Whether the relationship between these 'sub-elements' is essential, relevant and significant in constituting production, exchange and consumption are matters for further investigation.

11 I will use the term 'purpose' (and sometimes 'role') rather than the more technical term 'function'. However, it should be noted that by describing something according to its purpose, I am not attributing some teleological sense to it and implying that it is designed or created for this purpose. Rather, because of the way it is designed and created, it can be used, adapted and moulded for some higher purpose, in the constitution of something else. In Chapter 7, I will suggest that we exercise power by insisting that housing has a use (a purpose or role) that accords with our understanding and interests. Indeed, conflicts around housing have their grounds not in what housing is but rather in the differing uses to which it can be put. It is around these uses that a dominant ideology is formed.

12 On this ordered set of relationships as a hierarchy of mutual interaction see Michael Shute's (1994) article, 'Emergent Probability and the Ecofeminist Critique of Hierarchy'. For a more developed and sophisticated elaboration of what Lonergan calls 'emergent probability', see Lonergan ([1957] 1992: Chapter 4 and Chapter 7) and Melchin (1999, 2003). Later in this chapter an indicative theory of rent-setting will illustrate how we can understand the particularity of rental systems in different countries as an ordered set of relationships.

13 One difficulty here is that most of these examples not only elaborate on a particular discipline but also go a step further and critique what is currently happening by revealing situations of oppression, sexism, racism, classism and imperialism (resulting in inequity, injustice, non-participation, disrespect and lack of compassion) as particular groups seek to maintain and promote their interests. In other words, they deal with different questions at the same time rather than separating them out.

14 To paraphrase Lonergan's view of theoretical understanding as seeking 'to solve problems, to erect syntheses, to embrace the universe in a single view' [1957] 1992: 442).

15 'Personal identity' is a general term to incorporate all aspects of the development of a person – biological, psychological, intellectual, moral and spiritual.

16 By defining disciplines according to a particular purpose, it restricts the imperialist tendencies of one or other discipline, for example, the tendency of economics to dominate politics and culture. As long as these disciplines seek to 'explain' economic, political or cultural events in terms of the motivations and attitudes of social agents,

they will be frustrated because these events are not systematically related to personal attributes such as motivations and attitudes.

17   These needs and desires need to be affirmed on the basis of a fuller phenomenology of human deciding and acting.

18   Thus, rather than the current focus on an 'undifferentiated' market, on the motivations of consumers, producers and investors and on prices, economics could outline the functional relations that constitute an economy and distinguish between two types of firms: the first utilises technology to produce machines and equipment (a surplus circuit of production); the second acquires and utilises these machines and equipment to produce goods and services (a basic circuit of production) and thus, a standard of living. See Lonergan [1942–1944] 1998 and [1983]1 999. Also Shute 2010a, 2010b, 2010c, 2010d), McShane (1980, 1998, 2002) and McNelis (2010).

19   Politics is about how actual agreements are reached in various situations whether through negotiation, reference to authority, reference to expertise, force, manipulation, benevolence, bribery etc. Politics is not simply about state government and the process whereby they are formed nor about formal governance within organisations.

20   Culture then is not simply what might be called 'high art'. Rather it is what informs a way of living, what makes that living meaningful. Most prominently, a culture varies between national, ethnographic and religious groups. But within any society it also varies from group to group, and the culture of one group may complement or conflict with that of another. So, the culture within working groups – different professions, managers, workers – will vary as does the culture between children, young people, adults and older persons, between urban and rural groups.

21   These are brief descriptive answers to the questions 'what is an economy?', 'what is a polity?', 'what is a culture?', 'what is personal identity?' These answers are by no means adequate. They are merely indicative. As discussed previously, the answer to each of these questions is a theory of economy, a theory of polity, a theory of culture, a theory of personal identity. A theory seeks to grasp the significant and relevant elements that constitute or bring about each of these.

22   The variables within each level are speculative. The purpose of Figure 5.1 is illustrative.

23   These are broad disciplinary categories, each of which is developed around a specific purpose. It should be noted, however, that internal to each discipline are many different purposes that relate to one another. For example, internal to economics is the purpose of money, the purpose of the basic circuit of production, the purpose of the surplus circuit of production and the purpose of redistribution; internal to culture are various meanings, each of which has a purpose in relation to the purpose of culture.

24   For an example of empirical work relating housing and politics, see the application of Esping-Andersen's (1990) typology of welfare state regimes to housing by Joris Hoekstra (2010) and Kath Hulse (2003).

25   These elements are loosely drawn from the structures of housing provision proposed by Ball and Harloe (Ball 1986; Ball & Harloe 1992; Ball 1998).

26   In speaking about the role of social housing, I am only indicating the possible roles it can play. What role it actually plays and how this came about, whether this actual role has been successful and whether this role should change to something else are further issues that still require investigation in the functional specialties History and Dialectic. A theory of rent provides a heuristic for these further investigations.

27   For this different theory of an economy see Note 18.

6   History and the historical question

Dialectic   Foundations

History

Policies

Interpretation   Systematics

Research   Communications

The history of any particular discipline is in fact the history of its development. But this development, which would be the theme of a history, is not something simple and straightforward but something that has occurred in a long series of various stages, errors, detours, and corrections. To the extent that the one studying this movement learns about this developmental process, one already possesses within oneself an instance of that development which took place perhaps over several centuries. This can happen only if one understands both the subject and the way he or she learned about it. Only then will one understand which elements in the historical developmental process had to be understood before the others, which were the causes of progress in understanding and what held it back, which elements really belong to the particular science and which do not, and which elements contain errors. Only then will one be able to tell at what point in the history of the subject there emerged new visions of the whole...

Bernard Lonergan in *Understanding and Method* ([1959] 2013: 175)

In his article on 'Historical Perspectives and Methodologies: Their Relevance for Housing Studies?', Keith Jacobs notes how housing research has begun to explore social science methodologies but has largely ignored 'historical research and its associated methodologies'. He also notes that

> most housing academics make reference to some history in their research. This often takes the form of a short discussion at the start in the form of an introduction to the main study. The purpose of these historical introductions is usually to set out the context for the research and provide a chronology but little else.

He goes on to ask: 'What then can the housing specialist learn from the perspectives and methods of the historian?' (2001: 127–128).

So, what is the role of history in housing research? Is it about setting the context for some piece of research? Is it simply something of interest that bears no relevance to present day concerns? Is it simply a morality tale from which we can learn what to do or what not to do, through which a society passes on lessons to future generations?

Or, is the understanding of history essential to meeting the current challenges facing housing? In this chapter, as I work towards a more precise understanding of History as a functional specialty, I will affirm that history is essential and the way in which it is essential. I will argue that history is more than a narrative of events.

The vector of change: an initial view of History

In the previous two chapters, I have noted that in asking questions housing researchers distinguish between data that they turn to in finding an answer to their questions, and an answer which is the meaning or significance of these data. But I would also note that over time these questions change: they may change as a research project develops; they may change as a researcher shifts from one project to another. While a single topic may remain an issue over a long period of time, the questions asked about that topic and the answers given to those questions may continually change. Indeed, the answers to a single question form a sequence, an ongoing series of answers. A researcher could ask why this sequence of answers changes. Is there some vector or pattern or trend in the sequence? What is the vector that shifts the researcher from one answer to another answer?

Each researcher has their own history, but so also does the topic they are investigating have its history. Housing is ever changing. Between one time and place and another, there is the sometimes slow and sometimes quick change whereby what was the form of housing in one country in the nineteenth or twentieth century becomes another form of housing in the same country in the twenty-first century. As a result of this series of particular ongoing changes, housing in one place will differ from that in another. Each country has its own history, its own sequence of particular forms of housing.

So in relation to housing researchers, the question arises as to whether a sequence of questions within a researcher is ad hoc and random or whether there is a vector, a pattern or trend in the changes. Similarly in relation to housing, the question arises as to whether a sequence of actually operating housing systems is ad hoc and random or whether there is a vector, a pattern or trend in the changes.

On an initial view, then, the functional specialty History seeks to understand this vector of change within a sequence of events whereby the housing researcher and actually operating housing systems change over time.

History within housing research

History is the genre of housing research that looks back at events in the near or distant past and traces the changes that have occurred. Some of these histories are written by specialist historians, some are scholarly works by authors whose primary interest is housing, many provide a context for policy discussion or for further research and some are popular histories.

Specialist historians are primarily interested in the history of housing rather than current policy issues. In the UK, for example, Martin Daunton has written on housing in Victorian London (1991) and on the demise of the private rental sector and the rise of owner-occupation (1987). He also brought together other historians to write on the emergence of council housing between the twentieth century wars (1984) and on workers' housing in the late nineteenth and early twentieth century in different European cities as well as cities in the United States (1990). Urban historian Alison Ravetz has also written on the history of council housing (2001).

In Australia, specialist historians have produced few works on housing, and most of these focus on aspects of social housing. For example, Susan Marsden's (1986) history of the South Australian Housing Trust (SAHT), *Business, Charity and Sentiment*, recounts the development of the Trust from its origins in 1936 through to 1986. It describes various aspects of the Trust's operations as they changed over time as well as its personnel, in particular, the activities of the General Managers and the Board. The history encompassed 'not only the growth and functions of the Trust itself but also social, political and economic developments at State and national level as they affected or were affected by public housing' (p.vi). It describes the initiatives of the Trust, the range of its activities (as a development authority (passim) and as a migration agent (p.235ff), among others) and the changes that made the SAHT what it became in the 1980s.

Mark Peel (1995) in *Good Times, Hard Times: The Past and the Future in Elizabeth* picks up on one development initiated by SAHT. He traces the history of Elizabeth, a small town on the outskirts of Adelaide, from its foundation through to the difficult times of the late 1980s and early 1990s. Peel, who grew up there, intersperses a history of Elizabeth as a planners' vision to create a new town drawing upon the English tradition of the garden cities with personal reminiscences of a good place and a good community. SAHT sought to develop a community in Elizabeth, in part by mixing public housing and owner-occupied housing. It was, however, a town which came to rely upon its industrial links to the automotive industry; with the downturn in that industry, the good times turned into hard times and, particularly in the view of outsiders, the vision of Elizabeth faded.

Renate Howe (1988a) brings together a mixture of specialist historians, academics and practitioners in her edited book on fifty years of public housing in Victoria. In the first half of the book, specialist historians trace early attempts at housing reform in Melbourne (Harris 1988), the establishment of the Housing Commission Victoria (HCV) in 1938 (Howe 1988b), the early years in the context of post-war planning in Australia (Howe 1988c), the ethos of the post-war years, 'We Only Build Houses' (Eather 1988) and the experiment in slum clearance and high-rise flats that dramatically changed public housing in Melbourne for decades (Tibbits 1988).

Stephen Merrett (1979), Patrick Dunleavy (1981), A.E. Holmans (1987), Michael Harloe (1995) and Peter Malpass (1990, 2000, 2005) in the UK and Michael Jones (1972), Cedric Pugh (1976), David Hayward (1996) and Patrick Troy (1992, 2000, 2012) in Australia are housing researchers who have undertaken detailed historical research on some aspect of housing.

In his classic work *The People's Home?: Social Rented Housing in Europe and America* Harloe (1995) traces the history of social housing in five European countries (Britain, West Germany, France, the Netherlands and Denmark) and the United States of America from the late nineteenth century through to the late twentieth century. Within the ambit of social housing, Harloe includes three models in particular: residual, mass and workers' cooperative. He breaks the history of social housing into four periods: as a temporary solution after World War I, during the Great Depression, the golden age of reconstruction and growth after World War II and the revival of residualism from the mid-1970s. In each period, Harloe explores the economic, social and political context of social housing, the housing solutions and the role of housing reformers, the working class and other groups in proposing or opposing housing solutions. Using historical analysis, he is seeking to explain why social housing did not become one of the welfare pillars of advanced capitalism:

> As well as examining some of the evidence, this work seeks to understand why housing, through programmes of social rented housing in particular, has not taken a place alongside other significantly decommodified aspects of social provision, such as health, education and income maintenance, as one of the central pillars of the welfare state. (p.2)

Harloe rejects the view that housing solutions were the result of working class struggle or the benevolence of housing reformers and/or the state. The trajectory of housing solutions differs from other welfare pillars because of the pre-eminent role that private property plays in advanced capitalist societies. As commodified forms of housing are the norm within these societies, social housing is a residual sector. It is only in abnormal circumstances – at times of social crisis due to extreme housing shortages – that social housing has been called upon to provide housing. As he notes in the introduction:

> the interesting question, which this book explores, does not concern why housing has been such a marginal component of the welfare state but rather why it has sometimes been provided through the agency of the state in a partially decommodified form. More specifically, the examination will centre not on the ways in which the state has supported market forms of housing provision, either private renting or home ownership, but when, how and why the state's housing activities have been directed towards supplanting or making good a lack of these market forms through programmes of social rented housing. (p.4)

While Harloe's work is historical, his primary aim is theoretical. The book weaves together historical empirical evidence from the selected countries and a comparison between them showing a common thread in their development of social housing to argue that, in advanced capitalist societies, social housing is a residual rather than a mass model of housing.

In Australia, the primary focus of historical housing research has been on the development of government housing policy, in particular, the Commonwealth-State Housing Agreement (CSHA) (Jones 1972; Pugh 1976; Carter 1980a, 1980b; Kemeny 1981, 1983; Berry 1988; Hayward 1996; Troy 1992, 2000, 2012; and Jacobs et al. 2010a). For example, Jones (1972) recounts the history of public housing policies from the establishment of SHAs in the late 1930s and early 1940s through to the late 1960s. He focuses on particular elements such as means testing, socio-economic and demographic characteristics of those housed, the role in slum clearance and residential development, promotion of owner-occupied housing, rent policy and the social effect of public housing. Jones concludes by proposing some alternatives to public housing before suggesting some areas for reform. Hayward (1996) briefly outlines the history of public housing. He notes that the CSHA is the starting point for most such histories (p.5), but seeks to cast a wider historical net. Rather than beginning with the first CSHA in 1945, he goes back to early attempts by governments to respond to housing problems from 1900. For Hayward, this earlier history reveals a preference for owner-occupied housing, explains why it took governments so long (and very reluctantly) to turn to public housing as a solution to these problems, and contradicts the view put by Kemeny (1983) that the first CSHA 'coincides with the golden era in the history of public housing' (p.5). Hayward breaks the history of public housing into four phases, each initially characterised by the 'economic, political and housing context' followed by the policy response of governments, the changes in public housing policy (targeting, allocations and rent-setting) and the changes in the wider role of public housing in industrial, political and social development. He concludes that '[t]he twentieth century will end in much the same way as it began, with Australian Governments continuing to be reluctant, rather than willing, landlords' (p.32).

As mentioned previously, Jacobs notes that many housing researchers use a brief history to set the context for the discussion of a current issue. For example, McNelis (2000, 2004, 2006) precedes his analysis and discussion of the current role of social housing rental systems with a history of rental systems in Australia. Other authors focusing on housing provision for a particular population group provide a brief historical introduction: Watson (1988a) on women; Dargavel and Kendig (1986), Kewley (1973, 1980) and McNelis and Herbert (2003) on aged-specific housing; and Read (2000) (in an edited book) on Indigenous housing.

Popular histories such as the history of public housing in Canberra by Bruce Wright (2000) and the history of public housing in New South Wales by Housing NSW (2011) provide histories from the perspective a particular organisation.

What are these historians doing?

History as a genre of research tends to be relatively small with most historical research undertaken by those with an interest in housing policy. Their primary focus is to recount what happened over a period of time. As a narrative, these histories are concerned with individuals or governments who initiated some change in some particular time and place.

Second, these historians select certain events (decisions, activities and writings of particular people or particular groups or particular organisations) at particular times and places.

Third, these historians select certain events in view of some overarching theme which provides a perspective or way of pulling these events together. Events are selected and become significant because they are related to this overarching theme. Harloe (1995), for example, selects and relates certain economic, social and political events to social housing as a way of explaining why social housing takes a residual form rather than mass form in advanced capitalist societies. Jones (1972) is not interested in a history of public housing policies *per se*. Rather, his primary interest is in 'the attempts by governments to alleviate poverty' (p.vii). So, the history of public housing policies is outlined within this larger framework. His historical analysis reveals a number of 'real' problems: 'Most of the problems associated with bad housing…are not problems of housing in itself but of money; they are symptoms of low incomes' (vii–viii); the vagueness of the aims of public housing; and 'not lack of funds but inefficient use of existing resources' (p.vii). Thus, he selects events related to SHAs that are relevant to his overall theme. It is within such overarching themes that these historians trace changes over time and understand why they occurred or how we came to the current situation. This overarching view can be explicit or implicit. It can reflect a current rather than a contemporary viewpoint. It can reflect the biases and prejudices of a particular interest group.

Fourth, this historical research arranges events in a sequence and traces the transition from some past state of affairs to some later state of affairs. For example, Hayward (1996) arranges events at different moments in time (his four phases), notes the differences at different times and traces the transition in public housing policies from the early 1940s to the early 1990s. Similarly, Marsden (1986) arranges a sequence of events in the emergence of the SAHT and traces its development from early beginnings to what it became in the early 1980s.

Fifth, this historical research, particularly that by specialist historians, appeals to contemporary sources (such as autobiographies, newspaper reports, descriptions of witnesses, parliamentary inquiries and reports, Hansard and research reports) as evidence that certain events occurred. Each point in the narrative is supported by evidence that this event occurred in this time and place.

Sixth, for policy-oriented historians, the tracing of a contemporary policy issue over time sets a context for a current discussion. Past directions gives clues as to what might or might not work in the future, as to why something worked and something else didn't work, as to the difficulties that a particular policy might encounter.

Finally, it should be noted that, to different extents, these historians weave together empirical evidence, an overarching theme and a comparison between different countries or between different times continually moving from one question to another. For instance, Harloe is interested in and doing more than history. His primary interest is in making an assessment of the status of social housing within advanced capitalist societies and using a comparison of selected countries to reach some general conclusions.

Towards a more precise understanding of History

Is history scientific?

History as a discipline has been subject to some criticism, even within housing studies. For instance, in their discussion of the role of historical research in housing studies, Harloe (1995), Malpass (2000), Jacobs (2001) and Cole (2006) raise a number of issues: the tendency to view the past through the lens of the now; the tendency to produce accounts which inevitably lead to the current solution; the danger of over-simplifying the relationship between a problem and its solution; gauging the accuracy and reliability of sources; the role of subjectivity in historical research; and the use of current conceptual frameworks to understand the past.

These are complex issues and the extent to which these criticisms are applicable to the histories mentioned above is, no doubt, a matter of some debate. My interest here is moving towards a more precise understanding of History,[1] so I briefly discuss them in the context of a key question about history: do the techniques and achievements of these histories qualify them as a scientific approach? I suggest that they do not. Indeed, it is precisely because these historians (and their critics) have not made the radical shift to a scientific approach that viewing the past through the lens of now, the accuracy and reliability of sources, subjectivity, teleological accounts and conceptual frameworks become issues. The primary difficulty lies in the horizon of the historian, that of the taken-for-granted world of common-sense living, albeit with a more rigorous approach supported by evidence which has been assessed for its reliability and its validity.

As is common for housing researchers operating within the taken-for-granted world of common-sense living, historians (as historical researchers) do not distinguish between different questions: they do not distinguish between association-questions, what-is-it-questions, purpose-questions, change-questions, is-it-worthwhile-questions, what-to-do-questions etc. So their historical narratives move uncritically across these various questions. As a result, they operate across different functional specialties.

In the first place, as historical researchers seeking to determine whether particular events occurred at some time and place and whether these events are associated across time, they are operating within the functional specialty Research. As noted in Chapter 4, both use similar methods; yet the task of historical research is much more difficult because the data is more remote, its availability and extent is limited, it may come from a culture quite remote from that of the historian and it is more difficult to gauge its reliability and accuracy. A housing researcher, whether undertaking an investigation today or in the recent past or in the far distant past, still faces the same set of issues. Both the historical researcher and a housing researcher must deal with viewing the past (and the present) through the lens of now (out of their current culture), the tendency to produce accounts that accord with the views of their group whether social, cultural or religious, gauging the accuracy and reliability of sources and the role of their subjectivity in their research.

In the second place, a historian does not come with a blank mind, as the renowned British historian A.J.P. Taylor notes:

> History is not just a catalogue of events put in the right order like a railway timetable. History is a version of events. Between the events and the historian there is a constant interplay ... No historian starts with a blank mind as a jury is supposed to do. He does not go to documents or archives with a childlike innocence of mind and wait patently until they dictate conclusions to him. Quite the contrary.
>
> (Quoted in Jacobs 2001: 129)

While some of the 'historians' mentioned outline a framework of analysis or heuristic within which events are selected and understood, many do not. For instance, the histories of social housing organisations (Marsden 1986; Howe 1988a; Wright 2000) have neither a heuristic of organisation nor a heuristic of housing; histories of social housing (Pugh 1976; Carter 1980a, 1980b; Jacobs et al. 2010a) do not have an explicit heuristic of housing. For the most part, their heuristics are implicit. While Harloe refers to the structures of housing provision thesis, as noted in the previous chapter, the links between housing and its context are general.

Research, whether current or historical, presupposes some understanding of what they are investigating, an understanding that develops as their research progresses. Insofar as historians are developing an appropriate framework or heuristic within which to approach their historical research and within which to 'hold together' a viewpoint on this research, they are operating within the functional specialty Interpretation. As discussed in the preceding chapter, a theory provides such a framework by defining what is being investigated. Developing an adequate heuristic is the work of the functional specialty Interpretation. It is not the work of the functional specialty History.

If the work of the functional specialty History is neither (i) the work of the functional specialty Research determining the accuracy and reliability of sources in order to determine whether particular events occurred and in association with other events, nor (ii) the work of the functional specialty Interpretation developing an overarching framework (or conceptual framework) within which these events make sense, then what is the work of the functional specialty History? We are now left with a puzzling question. What is the distinguishing feature of History, what is its function?

History as a functional specialty

Jacobs (2001: 128), in answer to his own rhetorical question about what the housing specialist can learn from the perspectives and methods of the historian, replies: 'an historical approach provides us with the possibility of establishing trajectories or patterns within what might superficially appear as a disparate set of events'. The reference to 'trajectory' gets us some way towards a more precise understanding of the role of historians.

But this 'trajectory' has to be understood in relation to a what-is-it-question that has already been discussed in the previous chapter. We concluded by noting that a theory of housing is a set of related variable elements (according to Ball and Harloe these are production, exchange and consumption) whose range of possibilities is ordered by a range of higher purposes (what I have referred to more broadly as technology, economics, politics, culture and personal identity), each of which also admits of a range of possibilities. Thus, a theory is a complete ordered set of sets of related variable elements which will explain the particularities of any housing system in any time and any place.

If History is to have control over its research it requires an adequate theory of housing. To operate within History, housing researchers must shift into the world of theory where a theory of housing operates as a heuristic and anticipates the complete ordered set of sets of related variable elements that are relevant and significant to housing at any time or place.

What History pays attention to, however, is the actual operating housing system, a complete ordered set of sets of related *actual* elements. Indeed, History notes that there is a sequence of actual operating housing systems, and that these are continually on the move, with a new housing system emerging from the previous one. In a rather static fashion, Figure 6.1 illustrates a sequence of actual operating housing systems at points in time, Time 1 (T1), Time 2 (T2) and Time 3 (T3).

At a particular time in history T1, the actual operating housing system is constituted by the actual set of relations between its elements (production (P), exchange (E) and consumption (C)). The particularity of these elements is ordered by actual operating systems of know-how (technology), a standard of living (economy), a means of reaching effective agreement (politics), a common way of life (culture) and personal meanings and values (personal identity). By time T2, the actual operating housing system is constituted by a different actual set of relations between its elements (P_2, E_2 and C_2) in which the particularity of these elements is ordered by a different set of actual operating systems of technology, a standard of living, a means of reaching effective agreement, a common way of life and personal meanings and values. And by time T3, the actual operating housing system is constituted by another different actual set of related elements (P_3, E_3 and C_3) in which the particularity of these elements is ordered by other different sets of actual operating systems of technology, a standard of living, a means of reaching effective agreement, a common way of life and personal meanings and values.

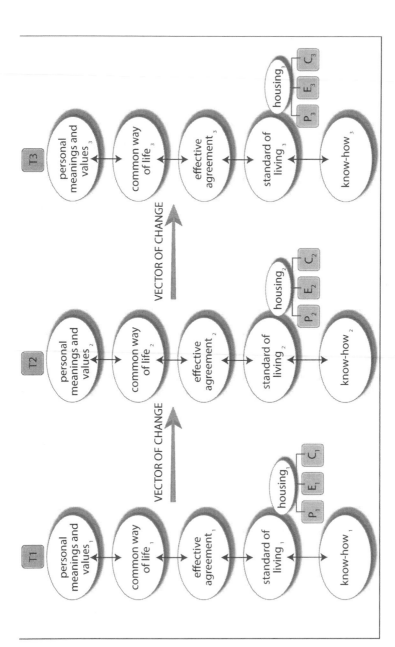

Figure 6.1 A sequence of actual operating housing systems

Note: The shadow on each box indicates the complete ordered set of sets of actual related elements that constitute the actual operating system for each purpose at a point in time (see Figure 5.2). The actual operating system for each purpose will change from one point in time to another.

At each point in time, T1, T2 and T3, the actual operating housing system can be explained in terms of the theory of housing as proposed in the previous chapter. As a heuristic, this theory identifies what is essential, relevant and significant; it distinguishes and relates various purposes to the ordering or systematising of the particularity of the set of related elements that constitute housing.

The discovery and documentation of each actual operating housing system is a task for a housing researcher operating within Research. Their investigation of different actual operating housing systems at different times and places provides data for better answering the what-is-it-question in Interpretation. The focus of History is on sequencing these actual operating systems in order to discover what brought about the shift from one to the next or, more precisely, what brought about the continuum of changes[2] (both small and large) such that an actual operating housing system at one point time 'morphs' into another at another point in time.[3] The task is neither one of describing or tracing the changes over a period of time, nor of comparing an actual operating system at two different times. Rather, the task is to discover the operative dynamic, the vector or principle that brings about the change in the actual operating housing system.

The goal of Interpretation is to develop a complete ordered set of sets of related variable elements that can explain all actual operating housing systems. Each purpose within the hierarchy of purposes is constituted by some set of related variable elements. Each purpose is realised through a set of related actual elements. Each orders or systematises the particularities of one or more of the related variable elements that constitute housing. So, a change in an actual operating housing system is the result of some change in the set of related actual elements that constitutes one of the purposes in the hierarchy. Where a change in the set of related actual elements is not simply a once-off but rather ongoing, it will make demands upon the particularities of lower level purposes (including housing). A change, then, in the actual operating housing system may result from introducing something new into: personal meanings and values (personal identify); a common way of life (culture); the means of reaching effective agreement (politics); a standard of living (economy); or know-how (technology).

These sets of related elements that constitute a purpose are sets of meaningful activities and processes. A change in these meaningful activities will have their origins within a particular person (whom the Historian may or may not be able to identify) and/or within a particular group. More significantly, as a group gives priority to achieving one purpose we could anticipate that this vector will become operative within their culture – it becomes a taken-for-granted way of operating within a culture. As such, it may clash with or complement other cultures, it may be common across a range of groups, it may become so deeply embedded in a culture that it becomes a presupposition for further decisions and actions and it becomes 'unnoticeable'.

A change in the actual operating housing system is predicated on some combination of the following (i) the discovery of new understandings in the set of related elements that constitute a purpose, (ii) the discovery that the current actual operating system (through which that purpose is realised) has its limitations, (iii) a drive to better realise that purpose, (iv) a decision by a group in co-operation with other groups that this realisation is worthwhile (and which changes the relative priority of this purpose) and (v) a decision by a group in co-operation with other groups to act upon these discoveries and to change some aspect of the actual operating system that realises this purpose.

A change, then, in an actual housing system may result from the discovery and implementation of: new personal meanings and values (personal identify) such as privacy and personal expression; a new common way of life (culture) such as the interdependence of people or respect for the environment; new means of reaching effective agreement (politics) such as consultation, participative democracy or tenancy legislation; a new standard of living (economy) such as aesthetically better dwellings; or new know-how (technology) such as better building materials, better construction techniques or payment facilities.

By considering not just one actual operating housing system at a particular time, but rather the sequence of actual operating systems, History is seeking to discover the vector of change, the operative dynamic that initiates the change from one system to another. It is an operative dynamic which, in some form, is likely to be still operative in the current actual operating housing system.

The theory of housing operates as a heuristic for understanding actual operating housing systems at different moments in time, for identifying the changes in these systems and for sequencing a series of systems. The theory of housing as a heuristic not only has these roles, however; it has a further role in the discovery of the dynamic operative in the sequence, the vector of change. Any change in an actual operating housing system will derive from any one of the purposes in which housing plays a role. The change cannot come from outside this set of purposes because the theory seeks to be all inclusive, it seeks to be a complete integrated viewpoint on housing.

The focus of the Historian is on this vector of change. This vector may be beneficial and lead to progress and development in the actual operating housing system. On the other hand, it may be detrimental and lead to a deterioration in the actual operating housing system as it begins to operate in the interests of a particular group. But such evaluations are not addressed within the functional specialty History. Rather, they are addressed in the functional specialty Dialectic, which is discussed in the following chapter.

To recap, the role of History is not simply to describe or trace the changes in a sequence of actual operating housing systems nor is it to compare the characteristics of actual operating housing systems within the sequence. Rather, its role is to grasp the dynamic that initiates the change in the sequence of actual operating housing systems, a dynamic that continues to operate over time and is likely to be still operative within the current actual housing system. An adequate theory of housing plays a heuristic role in two senses: the heuristic anticipates the complete ordered set of related variable elements that constitute housing at any time and any place; the heuristic also provides clues as to what may initiate the shift from one actual operating housing system to another and thence to another etc. by outlining all the roles that housing can play in the actual operating systems of other purposes.

Two indicative illustrations of History

Two examples seek to illustrate a more precise meaning of History: a discussion of social housing rental systems in Victoria (an Australian state); and a personal reflection on this researcher's developing understanding of rental systems.

Social housing rental systems in Victoria

Table 6.1 schematically outlines a sequence of actual social housing rental systems operating in Victoria since 1945. During this time, the sequence of systems is quite defined as they changed at particular moments in history. The heuristic for this sequence is the indicative theory of social housing rent outlined in Chapter 5. There, rent is constituted by three processes: gathering specified sets of data; specific operations carried out upon these specified sets of data resulting in a set of criteria for determining rent for each tenant; and the application of criteria to a specific tenancy. The theory also sets out the roles that the rent system plays: its role in the financial viability of social housing; its role in constituting the set of related elements that constitute social housing; its role in providing adequate housing that is affordable as one component of a standard of living; and its role in the economy.

In 1945 (Time point T_1), the then HCV adopted a dual rental system: a property rental system whereby the rent on a property is determined; and a household rental system whereby the rent for a household is determined according to their income. Table 6.1 outlines the respective datasets, operations and criteria that constitute this rental system. The property rental system, which can be described as a historic cost-rent system, used a dataset of the estimated historical costs of a dwelling or a project (a group of dwellings built at the same time); operated on this dataset by summing these estimated historical costs and divided the costs among all dwellings in a project; and applied the estimated costs per dwelling to a particular dwelling. The household rental system which can be described as an income-related rent system used two datasets: tenant incomes and basic wage; operated on these datasets with a rental formula; and applied the formula to the family income of a household. A household paid a household rent up to a maximum which was determined by the property rent.

In 1963 (Time period T_2), 1978 (Time period T_3), 1984 (Time period T_4) and 1992 (Time period T_5), HCV changed one or both rental systems. These changes, outlined in Table 6.1, involved some combination of changes in databases, the operations on these databases and the criteria used to determine the actual rent for a household. At each point in time, a range of purposes was operative in ordering the particularities of each rental system. The tension between two purposes can be highlighted: the financial viability of public housing in Victoria and its affordability for tenants. At the same time, the standard of living (in particular, the housing component) was changing; the individual and aggregated incomes of tenants were changing as broader economic conditions changed; and governments were adamant about the level and type of their commitment to public housing, as summed up in David Hayward's (1996) phrase 'reluctant landlords'.

Table 6.1 Rental systems in Victoria since 1945[1]

Type of rental system	Datasets	Operations	Criteria
1945–1963: T_1			
Property rental system (historic cost dwelling rent)	Estimated historical costs of dwelling or project (including debt repayment, maintenance, rates and taxes, insurance, vacancies and administration)	Estimated historical costs summed / With many dwellings in a project, costs were summed and divided equally between dwellings	Estimated costs applied to each dwelling
Household rental system	Tenant incomes / Basic wage	Rental formula[2]	Rental formula applied to family income
1963–1978: T_2			
Property rental system (historic cost pool rents)	Historic costs of all dwellings / Dwellings by type (house/flat) and size (number of bedrooms)	Historic costs summed / Number of dwellings by type and size determined / Relative weightings between type and size of dwellings determined / Rents for each type and size of dwelling determined by dividing the total historic costs among all dwellings with weightings for type and size	Type and size of dwelling determined and rent applied
Household rental system	As above	As above	As above
1978–1984: T_3			
Property rental system	Private rents by location, type and size of dwelling	Determine average private rent for location, type and size of dwelling / Determine and apply discount	Location, type and size of dwelling determined and rent applied
Household rental system	Minimum wage	Rental formula with (i) new rent-to-income ratio, (ii) new set of income included/excluded and (iii) different treatment of non-dependants	Rental formula applied to tenant/spouse income and non-dependants income and summed

Table 6.1 continued

Type of rental system	Datasets	Operations	Criteria
1984-1992: T_4			
Property rental system (current cost-rent system)	Current costs of administration, maintenance, rates and insurance Current values of all stock Rates of return on investment	Calculations of opportunity cost of investment (current value by rate of return) Depreciation calculated over life of dwelling Current costs summed Number of dwellings by type and size determined Relative weightings between type and size of dwellings determined Rents for each type and size of dwelling determined by dividing the total current costs among all dwellings with weightings for type and size	Location, type and size of dwelling determined and rent applied
Household rental system	No change	No change	No change
1992- : T_5			
Property rental system (market-derived rents)	Private rents by key characteristics (location, type and size of dwelling)	Dwelling rents sorted by key characteristics Dwelling rents averaged by key characteristics	Location, type and size of dwelling determined and rent criteria applied
Household rental system	Types of income received by households (according to source and/or purpose)	Inclusion/exclusion of income by source/purpose Determination of general rent to income ratio (ranges from 25% to 30% income) Specific determinations of ratio of income for different income types	Rent assessment formula (as outlined in rent manual) applied to particular income types received by household

Notes:

1. For a more detailed outline of public housing rental systems in Victoria, see McNelis (2000, 2004, 2006).
2. See McNelis (2000:48–49) for the details of this rental formula.

While the above arranges the sequence of social housing systems in Victoria since 1945, the question that remains to be answered is: what is the operative dynamic that shifted the social housing rental system from one actual operating rental system to another? Elsewhere, McNelis (2000) has argued that these changes stem from an ongoing instability in the social housing finance system as new tenants on lower incomes threatened the financial viability of HCV. In the face of the consistent refusal by governments 'to provide operational subsidies to SHAs to compensate for rental rebates provided to low-income tenants, and; the necessity that HCV (and other SHAs) maintain their financial viability' (p.116), HCV regularly underwent a financial crisis and changed the rental system to increase revenue, usually to the detriment of tenants. The operative dynamic here is the financial viability of HCV and this consistently had priority over public housing as a component of a standard of living (public housing standards reduced and housing affordability compromised).

A developing understanding of social housing rental systems

The previous illustration focused on particular changes in an Australian social housing rental system. One way of getting a sense of history, however, is by reflecting upon changes in oneself. The following is a personal reflection upon one part of my work as a researcher which has changed markedly over many years.

Housing research is constituted by series of processes that include asking questions about data, getting insights into this data or making discoveries, and affirming these discoveries. But the questions asked about data, the insights sought and the discoveries affirmed can change over time. And this is what happened to me as I sought to understand social housing rental systems. At three different times, I have expressed three different understanding of rental systems: in 2000 in a master's thesis, *Ideology and Public Housing Rental Systems: A Case Study of Public Housing in Victoria* (McNelis 2000)*;* in 2006 in an AHURI report, *Rental Systems in Australia and Overseas* (McNelis 2006; see also McNelis 2005); and in 2009–2010 as indicatively outlined in a paper delivered to the International Sociological Association International Housing Conference (McNelis 2009b) and in Chapter 5 above.

This work began in a context where others had sought to understand social housing rental systems. Their concerns were threefold: (i) the implementation and administration of rental systems (Victoria Ministry of Housing 1984; Ali 1985; Kleinman & Whitehead 1991; Yates *c.*1994), (ii) the resolution of issues raised by existing rental systems (Grey et al. 1981; Hills 1988) and (iii) understanding the context within which rental systems operated (Malpass 1990; Bramley 1991). My work built on this previous work.

The master's thesis began with a definition of a public housing rental system:

A public housing rental system consists of three parts: a specified group of dwellings managed by an organisation; a unifying rental principle which incorporates the values, goals and objectives of the organisation; and a rental structure which relates the rent on one dwelling to other dwellings.

(McNelis 2000: 4)

It then went on to describe various types of rental systems. It located them within a financial context and within some key principles of social housing (affordability, alleviation of poverty and equity), outlined the history of public housing rental systems in Victoria and concluded by evaluating Kemeny's (1995) theory of (rental) policy constructivism as an explanation for the way in which housing is structured in Australia.

In the 2006 AHURI report, the range of operating rental systems is widened to include many different systems in Australia and overseas. The report makes some changes in the types of rental systems. It more strongly locates the rental system within a social housing system, making some shift from understanding social housing and rents from differing points of view to understanding rent as it relates to (i) operating costs, (ii) the social housing finance system and (iii) the objectives of social housing (housing affordability, equity, workforce incentives and the autonomy of SHAs). Where the 2000 master's thesis noted these relationships, the 2006 report indicates more clearly how rental systems are related.

The 2009 paper distinguishes between two questions: what is rent? what purposes does rent play in the achievement of other purposes such as financial viability, affordability and social housing? It proposes a type of hierarchical relationship between the rental system and its purposes whereby: the possible elements and their relationships were particularised in such a way that rent played a role in achieving these higher purposes; the limits of this particularisation were the constitution of rent as a set of related elements.

These three are a sequence of expressions of understanding social housing rental systems. We can note the changes in the sequence: the 2000 master's thesis is largely descriptive of rental systems; the 2006 AHURI report shifts towards an explanatory understanding of rental systems; the 2009 paper and its fuller expression in Chapter 5 above is a more sophisticated explanatory understanding of rental systems as expounded in terms of a hierarchy of variables.

The key question for History is: what brought about the shift from one understanding to another? Or, to put it another way, what is the dynamic of change operating here? What is the principle of change?

The Historian can note the differences in questions asked as well as the differences in understandings of rental systems at each time and place. He/she can note a new or developing understanding of theory, the differentiation of theory from everyday common-sense living and the differentiation of theory from Research and from History (and so distinguishing theory as a hierarchy of variables from its role as a heuristic in understanding the actual operative process of history). However, it is not simply the desire to understand theory that is the dynamic that pushes forward changes. Rather, it is the decision to act on this new understanding, to realise theory as something worthwhile within research. It is the attempt to realise theory which initiates and pushes forward a new understanding of social housing rental systems.

The identification of theory as the dynamic operative through a sequence of understandings of social housing rental systems is an answer to the question 'what is the operative dynamic that brings about a change?' This can be distinguished from the question 'has this operative dynamic brought about a *development* in understanding social housing rental systems?' (a question for the functional specialty Dialectic) and from the question 'what do I do?' (a question for the implementation functional specialties).

Implications for housing research

Current historical housing research has primarily focused on Research, gathering and sequencing empirical data as it traces the transition from one state of affairs to another. Most of this research is undertaken within the context of an overarching theme that holds the narrative together. Some historical researchers use an explicit heuristic to trace changes in housing. The prime instance noted above is Harloe's use of the structures of housing provision to trace the history of social housing. For the most part, historians undertaking historical research lack an adequate heuristic through which they can control their investigation. They focus primarily on the motivations and attitudes of social agents. Yet, as previously discussed, these are but associations of time and place rather than systematic correlations. To some extent, these themes and heuristics are related to housing but the relationship is not specified.

The primary difficulty with current historical research lies in historians not grasping their primary function for much of their work lies within the functional specialties of Research and Interpretation. Through historical research, historians have significantly developed the techniques of Research and contributed to Interpretation. Nevertheless, they overlook their primary function; they do not seek to grasp the vector of history as history moves forward.

Figure 6.1 illustrated a sequence of actual operating housing systems as a hierarchy of purposes in which the particularities of the set of related elements that constitute an actual operative housing system are arranged in such a way that higher level purposes achieve their goal. Within this hierarchy of purposes, History identifies those purposes that have shifted an actual operative housing system at one moment in time to a different one at another moment in time.

A more precise understanding of history distinguishes between the theoretical question (what-is-it) and the historical question (what-is-moving-forward), a distinction between a constant and a variable. It is in history that different forms of the constant with its many particularities are created:

> A contemporary ontology would distinguish two components in concrete human reality: on the one hand, a constant, human nature; on the other hand, a variable, human historicity. Nature is given man at birth. Historicity is what

man makes of man…to understand men and their institutions we have to study their history. For it is in history that man's making of man occurs, that it progresses and regresses…

(Lonergan [1977] 1985: 170, 171).

It is in history that different housing systems are created. If we are to move forward in history and create a better housing system, we must not only understand what is essential, significant and relevant to constituting housing at any time and place, but we must also understand the dynamics of actual operating housing systems in the past, understand the dynamics which have led to the development of current forms of housing (tenure, design, type etc.).

There are many types of historical research. The challenge for historians, however, is to move beyond a concern with the events of the past and to grasp the dynamic of historical process. Their challenge and vital contribution to the future of humanity or, to use an old fashioned phrase in its more generic sense, to 'man's making of man', is to grasp the vector or dynamic operative in our current history.

The functional specialty History marks a paradigm shift in history as practised within housing research. It seeks an answer to a historical question, a what-is-going-forward-question, i.e. what dynamic or vector has provoked ongoing change in an actual operating housing system? The functional specialty History proposes a shift in history as currently practised in housing research *from* documenting and recounting a series of events in some remote or near past *to* using a theory of housing as a heuristic to grasp a sequence of actual operating housing systems and to identify the dynamic of change operative within this sequence, a dynamic which, in some form, is likely to be still operative in the current actual operating housing system.

Notes

1   This more precise understanding of History is loosely based on (i) Lonergan's presentation of 'genetic method' in Chapter 15 of *Insight* (Lonergan [1957] 1992), (ii) Chapter 10 in *Topics in Education* (Lonergan [1959] 1993) and (iii) Chapters 8 and 9 of *Method* (Lonergan [1972] 1990). Lonergan's presentation in *Method* tends to address a range of methodological issues within history rather than History as a functional specialty.
2   'Change' here is used in a neutral sense – it may result in development and progress or in stagnation and deterioration.
3   An analogy may give some sense of the difficulty faced by the Historian. We might compare their task with that of the botanist seeking to understand the dynamic or vector that transforms an acorn into an oak, or to that of the biologist seeking to understand the dynamic or vector that transforms a dog fetus into a fully grown adult dog. Just as there is an actual operating housing system changing over time, so too a dog is an actual operating system of integrated and inter-related biological systems (anatomical system, reproductive system, circulatory system, nutritional system, immunology system etc.) changing over time (a system on the move) from conception to life as a puppy, to life as an adult dog, to life as an old dog.

7   Dialectic and the evaluative/ critical question

Dialectic   Foundations

History   Policies

Interpretation   Systematics

Research   Communications

...in the world mediated by meaning and motivated by value, objectivity is simply the consequence of authentic subjectivity, of genuine attention, genuine intelligence, genuine reasonableness, genuine responsibility.

Bernard Lonergan in *Method in Theology* ([1972] 1990: 265)

Peter King begins his book *In Dwelling: Implacability, Exclusion and Acceptance* with a reflection (with reference to Derrida) on retracing our steps as a 'form of housekeeping', as coming to terms with what we have done and have achieved and as a way of understanding our journey:

No longer to like retracing our steps is to be against going home, rethinking or being prepared to start again. We are so convinced of where we are going that we take no precautions, we neither look back nor consider doubt worthy of us. We eschew the well-worn path that leads us back home. Whilst we might celebrate adventurousness and a positive attitude – to move forward – to retrace our steps actually demonstrates a preparedness to look back and to question where we are and how we got there; it is where we admit our mistakes, face the fact that we might have gone wrong – we are up a blind alley – and so we have not done all we could, not achieved what we thought we would, not found what we thought was there. But, of course, when we do retrace our steps we also face the prospect of unearthing what we have tried to hide, where we are so honest and open with ourselves that we have to accept we have failed, that we are lost, and so we might unsettle ourselves and perhaps others who depend upon us...We might see it is as a form of housekeeping...It is where we come to terms with what we have done and are therefore able to assess what we have achieved. Perhaps we should suggest that it is only by retreating to our base that we can fully understand the nature of our journey; that we can see it in all its consequentiality.

(King 2008: vii–viii)

The 'we' can be understood as ourselves personally or it can be understood as a group or society or culture. This chapter is about 'in dwelling', about ourselves personally and collectively as subjects in the world. It argues that by retracing our steps we can better grasp the dynamic of personal and collective development; we can retrieve and recycle the lessons of our personal and collective history. But such retracing is not just a personal quest. It is also a quest of our society and our culture. By reflecting upon this history, by retrieving and recycling what is worthwhile in our society and culture, we can create a history worth inheriting.

Retracing our steps: an initial view of Dialectic

In the last three chapters, I have used a phenomenology of research to distinguish initially each functional specialty. To date, this phenomenology of research has focused upon understanding. Now we need to advert to this understanding as an activity, as a human expression which proceeds from a decision of the researcher to pursue an answer to a question.

The pursuit of an answer to a question is one among many answers to many questions that a housing researcher could pursue. Its pursuit stems from a decision that this question is worthwhile, that it has priority over others. Indeed, research itself is but one among a vast range of possible activities that a person could undertake. For this person as a researcher, it is an activity that they have decided is worthwhile. It is, thus, an activity which takes some precedence or priority over other activities. Moreover, their decision to research some particular field or topic is a decision which provides the context for many subsequent decisions taken in pursuing this particular field or topic. These subsequent decisions presuppose this original decision.

On an initial view, Dialectic is concerned with past decisions and what motivated them. Dialectic is concerned with whether these past decisions are worthwhile or not. Dialectic reviews, assesses or evaluates past decisions as a priority over other possible decisions. Dialectic may take the form of a researcher reassessing their own past activities or assessing the activities of others. Dialectic 'retraces our steps' and assesses what we presupposed in our past decisions.

Retracing steps in housing research

The sociologist Loïc Wacquant[1] (2004: 97) notes

two senses of the notion of critique: a sense one could call *Kantian*…which refers to the evaluative examination of categories and forms of knowledge in order to determine their cognitive validity and value; and a *Marxian* sense, which trains the weapons of reason at socio-historical reality and sets itself the task of bringing to light the hidden forms of domination and exploitation which shape it so as to reveal by contrast the alternatives they thwart and exclude…

These two senses of 'critique' distinguish two genres of housing research that are largely oriented towards retracing our steps and considering past decisions. The first I will refer to as methodological critique which focuses on the presuppositions of researchers; the second I will refer to as socio-historical critique which focuses on housing policies and practices as an expression of some dominant culture that discriminates against or excludes, in some way, persons or households with specific characteristics (and give rise to various forms of ideology). Two other genres of research are also included: evaluation studies and comparative studies. They, like methodological critiques and socio-historical critiques, involve some sort of assessment or critique of past decisions.

Evaluation studies

Evaluation studies are predominantly focused on the decisions by a range of players that have resulted in housing projects, programs and policies having specific characteristics. Government or organisations undertake these studies to determine whether the originating objectives have been achieved, whether they have been achieved effectively and efficiently and whether and how they can make improvements.

Some evaluation studies are concerned with affordable housing projects, for example Ecumenical Housing et al. (1999) evaluate multi-unit developments in the ACT; Cloud and Roll (2011), the Denver Housing Authority's Park Avenue HOPE VI project; Milligan et al. (2007: Chapters 4 and 5) and KPMG (2005), the development of housing organisations in Australia; and the UK Department of Communities and Local Government (2010), the outcomes of mixed communities demonstration projects. Some are concerned with housing policy, for example, Boyne and Walker (1999) evaluate social housing reforms in England and Wales; Jacobs et al. (2010a; 2010b), public housing policy in Australia; and, Silverman and Patterson (2012), the Fair Housing Policy in the USA. Some are concerned with evaluations of neighbourhood renewal, for example, Hulse et al. (2004) evaluate the redevelopment of a public housing estate into a public-private neighbourhood.

Socio-historical critiques

While evaluation studies tend to take for granted an organisation's originating objectives for projects, programs and policies, socio-historical critiques are concerned with retracing the steps of more general movements within whole societies and cultures; they are concerned with classism, sexism, imperialism, bureaucratism etc., as they operate to distort human expressions and maintain the dominant power and position of certain groups within society. A critical stance is often taken for granted within housing research, whether it is being critical of government action or inaction, of housing policies, practices and projects, of how housing is understood, of political parties and ideologies, of different groups and their interests, of the terms used within housing research, of how research is being undertaken, of housing researchers etc. For the most part, the products of this critical stance are scattered throughout housing research.

Here, however, I want to highlight particular forms of socio-historical critique rather the incidental criticisms that arise randomly within housing research. One form of socio-historical critique is the way in which a dominant culture structures housing policy and housing provision in its own interests, excluding the interests of other parties. So, Kemeny (1981, 1983) and Ronald (2008) criticise the widespread acceptance of home-ownership as the best housing tenure and how its dominance has impacted on housing policies. Jacobs et al. (2010a) argue that the history of public housing and public housing policy in Australia 'cannot be understood without an understanding of the cultural basis of an ideology that sees private housing and owner occupation in particular, as a more desirable option'. As a result, public housing has developed 'a small but locally significant sector', one which is socially residualised and now placing more households at considerable 'risk of substantial affordability problems'.

Greg Marston (2000: 371) uses Critical Discourse Analysis (following Fairclough 1989, 1993, 1995) to examine changes in discourse between public housing managers and public housing tenants in Queensland as 'one site where ideological contestation and power are played out and legitimated'. In another article (2004) he explores the power relations within the Queensland SHA as senior managers sought to introduce a new managerialist approach. He notes the changes in discourse, the varying interpretations of managers and staff as they recognised the need for public housing reform (but not the excesses of managerialism), and the resistance of some staff as demonstrating that 'power relations may not be as fixed or unchangeable as they first appear' (p.18).

Jago Dodson (2007) examines the discourses on government housing policy to show 'the capacity of government to constitute and articulate the order of housing policy' (p.252). It is a critique that seeks to show the interests of governments and their bureaucracies in the formulation and implementation of housing policy.

Sophie Watson (1988a) in *Accommodating Inequality: Gender and Housing* critiques Australian housing policy for its exclusion of women's perspectives on housing. She examines how housing 'acts to both create and reproduce traditional family structures and the dependence of women' (p.viii). She begins with a historical perspective on housing women. In this history Watson takes the key events within the history of housing in the twentieth century and points to the assumptions regarding the place of women within these. She discusses how the Commonwealth Housing Commission (1944) and subsequent CSHAs implicitly understood the role of women within the family (pp.2–5). She shows how home ownership acted to exclude women and reinforce their dependence on their male partners and how their income and employment position, as well as discriminatory lending policies, made home ownership unlikely for women (pp.5–12): 'Woman… is constructed as dependent, as "housewife", as "homemaker"…Houses that are owned are perceived as providing the possibility for women to be creative, to blossom and to enhance the domestic ideal' (p.7). Watson then goes on to note how the resurgence of the feminist movement in the late 1960s and early 1970s put women's housing needs on the political agenda (pp.16–18). While women spend most of their time in the home, houses are not designed by them and are not designed to meet their needs (pp.18–20). She shows that housing policy and housing provision assume and promote the 'patriarchal family form'; home ownership excludes women; women in the private rental sector are on the margins of society; divorce and old age have detrimental housing implications for women; and a feminist analysis of housing has implications for urban theory.

Comparative studies

Comparative housing research is interested in how countries have developed different housing systems. It has become prominent over the past two decades presenting a range of challenges for housing researchers (Kemeny and Lowe 1998; Oxley 1991, 1989, 2001).

Some authors have used common analytic frameworks to compare countries. Peter Ambrose (1992) uses his 'housing provision chain model' to compare housing in Berkshire (Britain), Toulouse (France) and the E4 corridor of Stockholm (Sweden). Terry Burke (1993) uses a modified form of the 'structures of housing provision' to compare housing systems in the United States, New Zealand, United Kingdom, Germany, Denmark, Canada and Australia. Julie Lawson (2006) outlines the emergence of particular housing solutions in Australia and the Netherlands and seeks to explain their divergence by outlining how each emerged in the concrete circumstances of each country. She does so by elaborating on the necessary relations of housing provision within a changing environment, i.e. how these necessary relations of housing provision were actualised within contingent relations.

Tony Dalton (2009) compares the decline in importance of housing policy in Australia and Canada, arguing that the usual explanation for housing policy retrenchment, the shift to neo-liberalism, is inadequate. He proposes three causes: the interest of Keynesian economists in the production of housing rather than in its distribution; the limited capacity and marginalisation of state agencies responsible for housing; and the success or otherwise of civil society organisations.

In recent years, extensive work has compared Australia with other countries: Kath Hulse and Terry Burke (2005) compared social housing allocations systems across Australia, New Zealand, Canada, the Netherlands and the United Kingdom; Sean McNelis and Terry Burke (McNelis & Burke 2004, McNelis 2006) compared rental systems across Australia, New Zealand, selected European countries, Canada and the United States; Vivienne Milligan (2003) compared housing policies in Australia and the Netherlands and the housing affordability consequences for low income households; Nicole Gurran et al. (2007, 2008) compared practices in planning for affordable housing in Australia, Canada, USA, New Zealand, the Netherlands and Ireland; Jago Dodson (2006, 2007) compared government discourses on housing assistance policy in New Zealand, Australia, the United Kingdom and the Netherlands; Mike Berry et al. (2004) critically compared the ways in which Australia and the United Kingdom financed affordable housing; and Keith Jacobs et al. (2010a) compared the history of public housing policy in the United Kingdom and Australia.

Methodological critiques

Methodological critiques are the fourth genre of research that 'retrace steps'. This genre looks not to projects, programs, policies, nor the dominant interests within cultures or housing systems but rather to housing researchers themselves. In all genres of housing research, researchers evaluate/critique the methods used. Methodological critique raises questions about the adequacy of an approach to research, about the methods employed by housing researchers, about what their investigations focus on, about what is deemed important and what is deemed irrelevant and thus, to the conclusions they reach. Ultimately, these critiques are concerned with the housing researchers themselves.

The social constructionist and the critical realist are highly critical of the positivist tradition which pervades housing research. The social constructionist takes the perspectives and views of participants seriously, arguing that all meanings are social constructions (see, for example, Saugeres 1999; Jacobs & Manzi 2000; Jacobs et al. 2004). The critical realist argues for a reality beyond the constructions of participants. For instance, while Julie Lawson (2001, 2002, 2006) makes her critical realist starting point very explicit, she rejects positivism because it 'perceives the world as all that is observable' (2006: 14); she rejects interpretivism because it 'maintains that reality is defined by the meaning given by its inhabitants, rather than any objective, independent researcher' (p.15). She reviews alternative theories for the composition and dynamics of housing provision; first, describing a continuum of 'ontological alternatives' ranging from individual agency to structural explanations; and second, the different notions of causality within these 'ontological alternatives' and how they seek 'to explain difference and change in forms of housing provision' (p.58). In doing this she is seeking to show how these various alternatives 'fit' with a critical realist ontological framework which 'entails a structured notion of reality with related, overlapping domains of real (mechanisms), actual (events) and empirical (experience)' (p.45).

Jago Dodson (2007: 8) in his second chapter comments that 'there is no unified mode of inquiry that is specific to studies of housing policy problems, questions or issues' and that housing researchers 'tend to apply a variety of methodologies and theories'. He notes and critiques seven different methodologies: Marxist (Berry 1979, 1983, 1986), 'attenuated' Marxist (Ball 1983, 1986), the ideology of owner-occupation as developed by Kemeny (1981, 1983), libertarian (King 1996, 1998, 2003), feminist (Watson 1986, 1988b; Watson and Austerberry 1986), social constructionist and discourse analysis (Kemeny 1988; Jacobs & Manzi

1996; Clapham 1997; Batten 1999; Saugeres 1999, among others) and realist (Somerville and Bengtsson 2002). Dodson assesses each of these in terms of 'the extent to which they can address the problem of how to comprehend the connection between the construction of knowledge and empirical objects of housing, the determination of the character of the subjects of housing policy, and the means of deployment of knowledge of objects and subjects into empirical practices of housing policy' (2007: 23). He concludes that each of these methodological approaches is not adequate to the task and goes on in Chapter 3 to outline his own approach to housing policy, philosophical pragmatism.

Greg Marston (2002) argues for a new approach to housing research, critical discourse analysis. He reviews current approaches, including evidence-based approaches and social constructionist approaches. He illustrates critical discourse analysis, discusses its limitations but proposes that it is 'a valuable adjunct to the dominant positivist tradition' (p.83), one that 'can provide a valuable contribution to culturally and linguistically sensitive housing research' (p.92). In a wide-ranging assessment he concludes that housing research (along with other areas of social science) 'lacks reflexivity, creativity and originality' (p.189). He picks up Aristotle's distinction between techne (technical rationality) and phronesis (value rationality) (as developed by Flyvbjerg (2001)) and argues that housing research, particularly institutionally funded housing research in Australia, is dominated by technical rationality. He goes on to propose some principles for the renewal of housing research that 'places questions of power and value at its core'. He concludes: 'Without entirely throwing the rational evidence model of social research out, we need to pay more attention to those ethical and emotional means with which we may reframe political and housing policy debate' (p.189).

Jim Kemeny (1992), Chris Allen (2005, 2009), Tony Dalton (2009) as well as Rowland Atkinson and Keith Jacobs (2009) among others have noted how housing research has deliberately or unconsciously served vested interests. They have raised issues about the structure of the housing research industry, about the way in which problems are defined, about the quality of research, about the rise and fall in its fortunes, about the shift by governments towards contracting out research to accounting firms and consultants, and about the funding, organisation and politics of housing research.

What are these authors doing?

These four genres of housing research – evaluation studies, socio-historical critiques, comparative studies and methodological critiques – are concerned with 'retracing our steps', each dealing with housing in a different way: evaluation studies are primarily concerned with particular projects, programs or policies; socio-historical critiques are concerned with the presuppositions of dominating power (or their opposites, powerlessness or exploitation) demonstrated by various players – governments, managers, tenants, academics, policy-makers, practitioners etc. – through their actions and their discourses; comparative studies highlight the different trajectories of housing systems and their different emphases and preferences; and methodological critiques are concerned with the presuppositions of housing researchers.

All authors are gathering data. Depending upon the genre that they are operating in, the data they gather focuses on different aspects of housing: data relevant to particular housing projects, programs or policies; data relevant to the power relations between players in housing; data on the elements of housing systems; the writings of housing researchers. All authors are selecting and understanding the significance of this data within a particular interpretative framework. So, all are engaged in what we have previously referred to as Research and Interpretation. As such they are continually shifting between the empirical question, the theoretical question, the historical question and the evaluative/critical question.

The authors in these four genres look back at what has been produced: housing projects, programs and policies, dominant ideologies and discrimination against minority groups, housing systems within different countries and the methodologies used by housing researchers. In their different ways, they are all concerned with some state of affairs that has been produced as the result of past decisions and activities. This may originate from the decisions and actions of an individual, from a group or from a long history of a society or culture. Their primary interest is in coming to some judgement as to whether this state of affairs is good or bad. Common to all four genres is a view of science as emancipatory. In evaluating the worth or value of this state of affairs, they are bringing an ethical dimension into housing research. It is not enough just to know what is happening – it is also important to evaluate it critically, to reach some assessment as to whether it is good or bad (Flyvbjerg 2001; Marston 2008). In their evaluations and critiques, researchers implicitly sense that some state of affairs can be better than what it is. Those doing evaluation studies seek to learn from the experience of implementing housing projects, programs and policies and working out ways in which projects, programs and policies can be better in the future or ways in which they can be better implemented. Those doing socio-historical critiques seek to show that the current state of affairs is detrimental to a particular group and propose that a new alternative state of affairs should exist. By showing the way in which a state of affairs is constructed, socio-historical critiques of (taken-for-granted) dominant ideologies look to radical changes in the relations between dominant groups and those who are discriminated against. Those doing comparative housing research look for insights that will improve their understanding of housing; they also look for possible solutions to housing problems. While those doing methodological critique highlight the strong divisions between different methodological approaches and the inadequacies of these approaches, they are also trying to work out a better approach to housing research, one which will produce a better understanding of what is happening as well as better policies and courses of action to address housing problems.

Evaluation/critique is not simply about once-off particular decisions but rather about particular decisions as instances of a prevailing attitude or mentality. Evaluation/critique is concerned about the presuppositions or the framework within which these decisions are made. Evaluation studies are evaluating the practical frameworks that implement and produce a certain state of affairs; socio-historical critiques are evaluating the frameworks that are operative within different groups; comparative studies are evaluating the frameworks that are operative within a whole society; and methodological critiques are evaluating the frameworks that are operative within different types of housing research.

To effect their evaluation/critique, all housing researchers operating with these genres presuppose some framework of their own, one which differs from the framework of those whose decisions and activities produced these products. In evaluation studies, the researcher presupposes some view of what could be achieved and evaluates the actual achievement in terms of this. The researcher critiquing housing research as serving vested interests presupposes some larger view of housing research beyond solving short-term problems and the agenda of funders. In comparative studies, the researcher presupposes a framework which goes beyond a single country and encompasses all countries being compared. On one level, the comparison could be simply descriptive. A more thorough comparison would recognise that these descriptive differences are the result of different decisions in each country, decisions which give priority to different aspects of housing such as owner-occupied housing rather than rental housing, private rental housing rather than social housing, detached housing rather than units, housing affordability rather than appropriate housing etc.

While methodological critiques are often portrayed as philosophical disputes remote from the practicalities of undertaking research, differing positions on ontological, epistemological and cognitional issues operate to provide different and conflicting results. The researcher's framework provides a different perspective on the decisions and activities of others. It is a framework which selects or highlights certain events as significant and relevant, which selects a particular heuristic for understanding a situation, and which selects particular methods for going about research. While largely presupposed, this framework may to some extent be articulated. For instance, the evaluative framework for evaluation studies is usually the original stated aims and objectives of the project, policy or programme under consideration. It is against these that the mechanisms and outcomes of evaluation are assessed to determine whether they have or have not been achieved. But even here the adoption of these stated aims and objectives presupposes a perspective on how projects, programmes and policies are implemented.

Some housing researchers, as illustrated by Julie Lawson (2006), not only acknowledge different frameworks for evaluation/critique and outline their own framework but also seek to show how their framework incorporates other frameworks.

Towards a more precise understanding of Dialectic

While the primary interest of these genres of housing research is with an evaluation/ critique of some state of affairs, they also include components of gathering data (Research), interpreting its significance (Interpretation) and placing that significance within a short- or long-term history (History). I am proposing instead that the differing questions that underpin Research, Interpretation, History and evaluation/critique need to be separated as the methods used for answering each question are different. As I work towards a more precise understanding of the functional specialty Dialectic, my focus is on these studies specifically in relation to evaluation/critique.[2]

I also note that in running these separate questions together, these researchers indicate they are operating within an everyday common-sense framework. But this immediately suggests an alternative framework within which I am evaluating these genres of evaluation/critique and, a conflict between myself and the original researchers in this genre. As I retrace my own steps, it also includes a conflict within myself as I consider my own past research.

The problem of evaluation and critique

An article by Chris Allen (2009) in *Housing, Theory and Society* illustrates the problem of evaluation and critique within housing research. Allen questions whether knowledge acquired through formal research is a superior form of knowledge to that of ordinary people, to 'that which exists in the heads of people that live in houses'. He bases his challenge on epistemological grounds but his primary concern is that housing researchers are increasingly being used to justify government policies (p.53). The issue here is whether the understandings of housing researchers are any better than those of ordinary people. If they are not, then why engage in housing research at all?

Allen attacks the superiority of housing research to gather empirical data and to present conceptual frameworks that explain this data. He then goes on to raise the question of world-views and concludes that 'social sciences have no right to posit their own "world views" as superior and that, furthermore, the construction of a world view actually "falls outside the range" of the task of the social sciences' (p.70). With reference to a number of existential phenomenologists, he argues against the idea of a social science world-view. He quotes Heidegger (1988: 6) to the effect that a world-view 'always arises out of the particular factical existence of the human being' (p.71). He argues that for housing researchers, as for everyone else, their world-view and their understandings are a consequence of their facticity as beings, their being-in-the-world, their lived experiences and understandings. Thus, 'it makes no philosophical sense…to make a distinction between social science and local knowledge' (p.55).

In the course of his argument, Allen notes that housing researchers have not taken Jim Kemeny's work sufficiently seriously, in particular, quoting Kemeny (1992: 20): 'fundamental prior questions concerning the grounds of knowledge of housing studies: questions which have rarely, if ever, been addressed by housing researchers' (p.54). He is critical of housing researchers because they 'seem to have become increasingly *reflexive about the phenomena they are concerned with*. However, there is little evidence that housing researchers have become more *reflexive about the grounds of their own practices of knowledge gathering and production*' (p.54; emphasis is in the original). He goes on to argue that 'the notion of a "Housing Studies" is an epistemological fallacy, that is to say, that the idea of "Housing Studies", as it is understood by "housing researchers", is untenable in philosophical and epistemological terms'.

In the introductory editorial to the same issue of *Housing, Theory and Society*, David Clapham (2009) rejects Allen's position because it would mean the end of housing studies, indeed, of all social science. However, he does not reject it outright: 'Allen's critique is relevant to some forms of housing research but not all'. He argues that it is not relevant to the 'form of research which builds the conceptual understanding from the common sense worlds of ordinary people' (pp.6–7). He also notes that Allen is concerned with the viewpoint of but one group of players: current residents. Clapham rejects the role of housing researchers as the 'neutral arbiter of facts' because 'there are no neutral social facts'.

Clapham then goes on to propose a role for housing researchers and in doing so he shifts to a new context, that of the 'decision making process and the role of research and academics within it' (p.7). He views this process 'as a power game between competing groups with different "world views" or realities pursuing their own aims and interests' (p.8).

In this way, Clapham maintains some role, albeit limited, for housing studies. He proposes that one role is 'for research to describe the views of the different participants and to compare them using an analytical framework' (p.8). What Clapham seems to be doing is absolving the housing researcher from taking any position with regard to these differing world-views. Rather, their role is simply to present positions by describing and comparing them. Moreover, in so doing it is possible for housing research 'to be emancipatory in highlighting perceptions and views that otherwise may be hidden'. Clapham's analytic framework, however, presupposes some position or world-view in which priority is implicitly given to some rather than other characteristics that are compared and in the arrangement of those characteristics. His solution, then, doesn't really deal with the problem posed by Allen. Neither Clapham's analytic framework nor his suggested role for housing researchers (as the 'neutral arbiter of world-views') are neutral.

Both Allen and Clapham reject the empiricist world-view of objectivity and neutrality. They recognise that housing researchers cannot escape their own world-view and that it will continually inform their understandings, their decisions and their actions. Both are critical of some group: Allen is critical of those housing researchers who regard their world-view as superior to current residents; Clapham is critical of all housing researchers insofar as he proposes an emancipatory role for them.

However, both these critiques imply that there is some legitimate framework for housing researchers, for individuals and groups. Some may argue that there is no legitimate framework and that critique is merely arbitrary, simply a matter of our inheritance, of the family, group and culture into which we were born. But if this is so, how are we to address the issues of dominating power and self-aggrandisement? In the wake of the masters of suspicion[3] – Freud, Marx and Nietzsche – we can no longer assume that we are but innocent bystanders. That in most situations there is some exercise of dominating power and self-aggrandisement seems to me to be incontestable – indeed, it is to be expected. That many sociological studies are devoted to revealing the exercise of dominating power between people and between groups in very many differing situations does not take us far in our understanding of what is happening. Dominating power and self-aggrandisement are to varying extents exercised through all forms of human expression, whether practices, language, bodily movement, drama, social systems, culture etc. The core problem is grasping the way in which these varying forms have been distorted in the interests of a particular person, group, culture or religion. It is only in this way that we can propose policies that change a form of expression, or change the actual activities that constitute housing. If the issue is simply the exercise of dominating power and self-aggrandisement, the only solution is a counter-exercise of dominating power and self-aggrandisement by another person, group, culture or religion. Ultimately it is a matter of eliminating persons, groups, cultures and religions. Thus, the core problem for housing is one of grasping how activities that constitute housing are distorted by dominating power and self-aggrandisement as a prelude to proposing new directions for constituting housing. The first issue, then, is one of the appropriate framework that will inform the decisions and actions of individuals and groups (and get us beyond the distortions of dominating power and self-aggrandisement).

Allen is not simply critical of some group – he is critical of housing researchers. For him, the root of the problem lies in their understanding (and practice). Their approach to research, as much as it is dressed up in methodical rigour and analytic frameworks, does not differ substantially from that of ordinary people. Housing researchers may or may not share interests in common with particular groups such as local residents. Insofar as they do not share such interests, conflicts arise between researchers and particular interest groups, and between researchers in the conclusions they reach. Allen clearly recognises the limitations of current housing research and seems unable to envisage an alternative framework for conducting such research. The second issue, then, is one of the appropriate framework that will inform the decisions and actions of housing researchers.

But there is a third issue here. Allen and those housing researchers undertaking evaluation, socio-historical critique, comparative studies and methodological critique themselves presuppose a prior framework within which they operate. Further, those researchers undertaking an evaluation of those researchers undertaking evaluation, socio-historical critique, comparative studies and methodological critique also presuppose a prior framework as do those evaluating these researchers. And so on.

As noted above by Allen, Kemeny (1992) calls for greater reflexivity among housing researchers. He asks them to become aware of the implicit framework (episteme) they are using to understand housing. Here I am extending this call to include awareness of not only the implicit frameworks that housing researchers use to understand housing, but also the frameworks of housing practitioners and the frameworks of those housing researchers critiquing the frameworks of practitioners, of other housing researchers and of those doing the critiques.

Housing researchers not only face the issue of the best framework within which to critique dominating power and self-aggrandisement and the issue of the best framework within which to conduct their research, but also the issue of the best framework for critiquing housing researchers. Without a discussion of this issue, how can we move beyond the often bitter and prolonged conflicts between positivists, social constructionists and critical realists? How are we to come to a view as to the best way forward for housing research?

If we are to address these three issues, we need some method (which here is called Dialectic) to reflect on the past, search out the best of the past to come to a view as to which position produces progress; indeed, to come to a view as to what is progress. Dialectic evaluates the past. It seeks to acknowledge the achievements of the past, to critique its inadequacies (whether in everyday living or of housing researchers or those evaluating housing researchers) and to resolve past conflicts by reaching for an integrated view that will consolidate the worthwhile achievements of the past. Only then can we move on to realise something better.

I have introduced a threefold problem for evaluation and critique and also proposed a goal and program for the functional specialty Dialectic. As a preliminary to a more precise understanding of Dialectic, I now seek to deal with these complex issues. First, I expand the notion of theory by asking the question: what is theory? I then relate theory to values. Following this I will discuss the notion of horizon and distinguish three types: horizons of everyday common-sense living, horizons of theory and the horizon of the subject as subject. Finally I discuss Dialectic as a method for searching out the best of the past.

Theory, an expansion

In Chapter 5 I outlined some of the characteristics of theory: it is an explanatory definition that grasps the significant, relevant, essential elements and their relations that constitute something such as housing; it is abstract, universal, invariant and normative; it can function as a heuristic in other functional specialties.

In proposing a theory of housing, I presupposed a particular understanding of theory. This is one understanding among many and before we move on to a more precise understanding of the evaluative/critical question, I need to expand further on this understanding of theory. I do so under two headings: 'what is theory?' and 'theory and value'. The significance of this expansion will become clearer as I discuss the notion of horizon and dialectic as searching out the best of the past.

What is theory?

In Chapter 5 I proposed a theory of housing as the complete ordered set of sets of related variable elements that constitute housing and its particularities. Each set specifies the conditions for the achievement of a purpose within which housing plays a role. At the same time, each purpose orders and systematises some aspect of the set of related elements that constitute housing.

In answering the question 'what is theory?', we face a similar problem to that in Chapter 5. Within a scientific context, our orientation is towards asking and answering questions. It is a context within which we distinguish different types of questions and the answers they anticipate. Both questions, 'what is housing?' and 'what is theory?', anticipate the same type of answer. 'Theory' names something that, in some sense, we already know and the word 'theory' operates as a heuristic. The question 'what is theory?' presupposes occurrences of what we name 'theory'. We understand something of theory but, in our puzzlement, we are seeking a better understanding.

We could answer the question by examining the work of social and housing researchers and the theories they are developing under such generic titles as functionalism (Parsons [1951] 1991), grounded theory (Browne and Courtney 2005), rational choice theory (Bengtsson 1998), social constructionist theory (Jacobs et al. 2004) and critical realism (Lawson 2001, 2002, 2006). Or, we could answer it by examining the work of metatheorists, those who examine, compare and critically assess various social theories (for example, Sayer 2010; Ritzer 2003, 2005). From this menu, a researcher could make a leap of faith and adopt one or other approach, even vary it according to circumstances. This could fit with their ideological preferences or with the fashion of the moment. In doing so, however, it is doubtful whether the researcher has come to some understanding of what theory is.

If a researcher were not to make a leap of faith, on what basis would they adopt one or other approach to theory? On what basis would they be prepared to accept what someone says about theory – this is theory and that is not theory? A rigorous scientific approach would demand that a researcher at least sort through the different and sometimes conflicting activities of social scientists which they and others regard as theorising; that they sort through the various and sometimes conflicting positions on theory, the disagreements and arguments about what it is; that they bridge the gap between what a social scientist is doing when they are theorising and their understanding and articulation of this. This is the empirical data in which possible occurrences of theory are associated in time and place with other events/processes. It is *somehow* within this manifold of diverse and apparently ambivalent data that a researcher can reach some understanding of what theory is and make a discovery as to what it is. It is upon this understanding that a researcher can make their own judgement that this is theory and that is not theory.[4]

A key word here is 'somehow'. It is only in coming to terms with this 'somehow' that we have a sound basis for reaching a judgement that this is theory and that is not theory. In the discussion of the theory of housing, I distinguished between a descriptive definition and an explanatory definition. A descriptive definition of theory will not adequately distinguish between what theory is and the purposes that have become associated with theory. Our challenge is to develop an explanatory definition of theory, one that critically distinguishes between what is relevant to answering the question and what is merely associated (in time and place).

When I sought to develop a theory of housing I brought a heuristic (a current understanding of housing), I turned to a manifold of external events, I used 'fantasy and lateral thinking' to reach for the set of relations between elements that were relevant, significant and essential to the constitution of housing. Through insight, I grasped a theory of housing as the systematic correlations between elements, the complete set of related elements that constituted housing. I then went on to affirm that this was my best understanding.

In seeking an answer to the question 'what is theory?', we follow a similar process. The relevant data, however, are not external; rather, they are internal to a researcher. They are reached by a researcher turning inwards to the processes going on within themselves as they ask and seek an answer to a what-is-it-question. The question 'what is theory?' is an invitation to increased reflexivity (Kemeny 1992: xvii and Chapter 2), an invitation to each of us to undertake a personal experiment by asking the question 'what I am doing when I am theorising?' and paying attention to what is going on when we theorise.

The primary source of data about theory is within ourselves as we theorise. It is these data which will ground our understanding and stance on theory. It is an empirical approach in that it appeals to data. This internal data is attended to and affirmed only through a personal experiment. The data here are the activities of seeing, hearing, feeling, tasting, smelling, thinking, considering, supposing, defining, judging, deciding, weighing up, imagining, formulating, deliberating, inquiring, asking questions, speaking, writing etc. Within this array, we are seeking an insight into the complete set of related elements that constitute an explanatory definition, a theory.

I am proposing here that theory is constituted by the relations between three elements: (i) asking a specific question, a what-is-it-question, about some data or events associated with one another in time and place, (ii) grasping, through an insight, the systematic correlation between a set of related elements whose occurrence conditions the occurrence of the 'what' and (iii) verifying or affirming this insight as the best possible. Theory, then, is constituted by 'an integrated set of insights (more or less probably verified) and questions' (Melchin 1999: 11). This set of related elements is an explanatory definition of theory, a theory of theory. Such a definition is unusual because, unlike others which highlight the material basis of theory, it grasps the related elements operative within a researcher.[5]

This theory of theory is universal, invariant and normative – it grasps the complete set of related elements (a what-is-it-question about data, a set of insights that grasp a systematic correlation between elements that constitute this 'what', and an affirmation) that constitute theory in whatever time or place it occurs.

The introduction to Chapter 5 outlined a dispute between those who proposed that the role of housing research is to *apply* social theory to a housing issue and those who proposed that its role is to *create* a housing theory. As a way of untangling some of these messy debates, in the conclusion to Chapter 5 I argued that theory plays both roles. A theory of theory relates a what-is-it-question about data, a set of insights that grasp systematic correlations between a set of elements and an affirmation. As a heuristic, we take a theory of theory and apply it to some particular research. As such it is a heuristic, a guide to creating a theory of housing, a guide to answering any what-is-it-question. So, an answer to the questions 'What is housing?', 'What is research?' will be a set of insights that grasp systematic correlations between the set of elements that constitute housing and research. As a heuristic, it applies a theory of theory in order to create a theory of housing and a theory of research.[6]

Further, as discussed in Chapter 5, theory occurs within a context. Its occurrence is conditional upon the occurrence of a set of related elements; the occurrence of this set of related elements is also conditional. So, we can ask about the conditions for the occurrence of each of these elements and these conditions will be understood as answers to such questions as 'What is data?', 'What is insight?', 'What is affirmation?' So, data depend upon our capacity to see, hear, feel, imagine. They also depend upon our consciousness of ourselves as operating subjects. Insights depend upon our capacity and our willingness to ask and to follow through the relevant questions and on our capacity for 'fantasy and lateral thinking'. Judgements depend upon our capacity and willingness to seek what is really happening and not just some possibility. But my topic here is not to elaborate further on a phenomenology of inquiry and the conditions for data, insights and affirmations. I do, however, want to note that theory occurs where its conditions are fulfilled and that these conditions themselves occur where their conditions are fulfilled.

Asking questions, making discoveries and affirming discoveries is not restricted to theory. Rather, it is common to all forms of thinking and evident in all facets of living, in the everyday taken-for-granted world of common-sense living, in the world of aesthetics, in the world of mathematics etc. Our explanatory definition of theory distinguishes theory from other forms of thinking. Just as a theory of theory is a higher viewpoint which systematically relates a what-is-it-question about data, insights and affirmations, so too a theory of data is a higher viewpoint which systematically relates seeing, hearing, feeling and imagining. Further, just as a theory of theory is a higher viewpoint which systematically relates a what-is-it-question about data, insights and affirmations, so too theory itself can play a role and become one of a number of conditions for a higher viewpoint that we have called 'science' which in turn can play a role and become one of a number of conditions for a higher viewpoint that we might call 'human living'. Just as within the theory of housing we proposed an expanding series of contexts, so too there is an expanding series of contexts in regard to a theory of theory.

Theory and value

A theory of housing is developed retrospectively as a reflection upon what has occurred. We can, however, think of it prospectively. Housing will be created when the set of related elements occur. Prospectively, we create or construct housing because it is something worthwhile, it is something of value. Its origins lie in the decisions of individuals and groups and when we deliberate about a course of action we are *intending* something more than what merely satisfies or pleases us, something more than what is apparently good – we are intending what is worthwhile, we are intending value. So, when we decide to bring about housing, we are constructing something that we have decided is worthwhile. A theory of housing, then, grasps the conditions required for the occurrence of housing as something worthwhile, something of value (Lonergan [1972] 1990: Chapters 1 and 2). This understanding of 'value' differs from the usual understanding where housing researchers refer to 'their values' or politicians and community leaders refer to 'family values' or 'organisational values'. Statements do not reveal values, decisions and activities do.

As proposed in Chapter 5, the conditions for the occurrence of each 'what' – whether a standard of living, effective agreement, a common way of life, personal meanings and identity – is a complete set of related elements. Each 'what' is created or constructed as something worthwhile, as something of value through the decision of an individual or group and the sets of related activities that proceed from that decision. As Kenneth Melchin notes (1994: 25) in an article in which he explores the relationship between economic and social structures and ethics:

> the economic explanations themselves articulate forms of human co-operation toward goals and it is this dynamic relationship between co-operative structure and goal which is the essential meaning of 'value' in the social sense. The explanandum of economics is 'value.' To achieve their objectives, economists need to recognize this fact and continually search out heuristic and explanatory tools which are appropriate to understanding co-operative social schemes.[7]

While a theory of housing grasps the set of related elements for a value to occur, the ordered set of theories grasps the role that housing plays in constituting other values and so locates it within a hierarchy of values. As I noted previously, a particular type of hierarchy is operating here, not one where the higher dominates the lower, but rather one of mutual dependence in which: each value has its own autonomy in that its occurrence depends upon the occurrence of a set of related variable elements; a higher value depends upon the occurrence of a lower value and is limited by the range of possibilities within the set of relations in the lower value; the higher value orders or systematises the elements in the lower value in such a way that the particularities of the lower value facilitate the occurrence of the higher value. The theory of housing, then, outlines a hierarchy of values, but it also outlines the mutual dependencies between housing and those other values in which it plays a role.

But there is a further point to note here. While housing is a value intended and the theory of housing grasps that it is constituted through a set of related variable elements, housing is actually realised/created/constructed through an actually occurring set of related elements. While housing is a value intended, the outcome realised through an actually occurring set of related elements may be limited or inadequate in some way. As a value intended, housing becomes the criteria for criticising the actual outcome and for criticising the actually occurring set of related elements that produced this outcome – both the outcome and the processes could be better than what they are. We can therefore distinguish between the particular good that is realised, the actually occurring set of related elements that realised this good, and the value whose realisation we are seeking. It is the attempt to better realise this value that motivates changes in the actually occurring set of related elements.

Horizons

Housing research is characterised by ongoing conflicts. Such conflicts can arise as a result of differences in the data that is attended to. These can be overcome by gathering further data. Conflicts can also arise as a result of differences from asking different questions, from incomplete understandings and from starting at different places and times. They result in different perspectives that point to the complexity of our world (Lonergan [1972] 1990: 214ff, 235). Some differences, however, are more fundamental. They regard the fundamental frameworks that inform our way of life and our work as researchers. Conflicts in fundamental frameworks are much more difficult to resolve.

For each of us, a background framework is the set of pervading presuppositions that inform all our decisions. These are implicit within the culture of any group, whether it is an informal group, a small organisation, an ethnic group, a religious group, a linguistic group etc. Each business group, industry group, bureaucracy, social institution, artistic group, academic discipline, research group operates with some implicit presuppositions. It is what holds the group together. They have developed over time in response to particular situations. They change slowly.

Every inquiry has its presuppositions whether researchers know it or not, whether they explicitly acknowledged their presuppositions or not, whether the researcher is operating within any of the functional specialties (Lonergan [1972] 1990: 247). These presuppositions of living and inquiry are referred to as a 'horizon'.[8]

Horizon is the line that separates the earth from the sky. But here it is used in an analogous sense of what separates the totality of each person's concerns from what is outside those concerns: 'Within one's horizon, one's ready-made world, one is organized, one has determinate modes of living, feeling, thinking, judging, desiring, fearing, willing, deliberating, choosing' (Lonergan [1959] 1993: 90). Horizons are 'the sweep of our interests and of our knowledge; they are the fertile source of further knowledge and care' (Lonergan [1972] 1990: 237).

Our horizon defines our world, what we attend to and what we are interested in and care about. It orients us in our world. Horizons develop slowly as we live, as we learn: the scope of what we understand and can understand advances; the range of our skills and the potential for expansion develops further; the range of situations (technical, social, political and cultural) we can deal with, first with difficulty and then easily, expands.[9]

Horizons of everyday common-sense living

Everyday living is oriented towards solving the many and various immediate problems that can arise as we live our lives (among others see Goffman 1961, 1969; Garfinkel 1962, 1967; Schutz 1972; see also Lonergan [1957] 1992: Chapter 7). Through our socialisation or acculturation, we learn the ways in which people in the past have dealt with these many problems. It is this learning from the millennia of human experience that we carry within us as we work out how to deal with the situations that confront us. This learning is formed by the ongoing cycle of particular problems and particular solutions. It is this history of particular problems and solutions that gives rise to different cultures, different mores and different economic, political, educational, legal and other institutions.

Within this horizon of everyday common-sense living, our orientation is decidedly towards what is immediately required. Our understanding is oriented towards *understanding things as they relate to us* (or the group we belong to). So while we feel the demand to do what is best, or feel the demand to improve what is happening, we find we do not want to or do not have the time to pursue our questions about what to do beyond the solution we need for our current problem. We cut off our pursuit of questions because our living demands immediate responses. Our learning is through trial and error, trying things out and discovering what works and what doesn't. It is this immediate orientation that puts in place and continually adjusts the vast structures of economy and society, health and education, housing and urban living, knowledge and technology, history and culture, literature and the arts etc.

But this cutting off of questions limits the extent to which we, operating within the everyday world of common-sense living, can understand our social situation. The limits of our horizon come to the fore in our encounters with other people: those we are intimate with; families and friends; the groups in which we live, work, learn and recreate; and people from other similar or alien cultures. Here we find that other people understand things differently, and have different and sometimes unsettling interests and concerns – what we regard as vital and important they may regard as secondary, irrelevant and inappropriate. They may screen out altogether what we regard as important, and vice versa. Indeed, a decision to notice, to understand, to talk about differences is a decision to shift out of our own narrow horizon to a larger one that encompasses the other person; that is willing to see things from their point of view; and that is willing to recognise that other horizons of everyday common-sense living are legitimate (Melchin 1998).

Besides the gradual – at times slow, at other times quick – development of a horizon, there can be the dramatic and radical transformations of a horizon. The classical examples are: falling in love and the discovery of being loved where what was previously regarded as important no longer is and what was previously ignored or ridiculed becomes vitally important; or the transformation of Helen Keller as she discovered that the movements in her hand by her teacher, Anne Sullivan, meant 'water'; or the discovery of a life-long passion, whether an artistic pursuit, a social or political pursuit or an intellectual pursuit.

But we may decide not to notice. We may be incapable or unwilling to understand things from another point of view or we may reject that point of view. So conflict arises between individuals, between groups, between cultures, between religions. Conflicts have become the 'grist for the mill' of social scientists: studies of classism (such as Marx's *Das Kapital*) document the conflict between the bourgeoisie and the workers and build an understanding of an economy and society on this basis; studies of cultures (such as Nietzsche's 'will to power') document the motivation and justification for one group (with their particular horizon) to dominate another (with their particular horizon); studies of imperialism document attempts by one country to dominate another; studies of racism document attempts by one race to dominate another; studies of sexism document the dominance of patriarchy; studies of militarism document attempts by military means to dominate a country, ethnic group or civil group etc.

Each person and each economic, social, political, ethnic, national, cultural and religious group has their own horizon within which they operate, understand the world, solve the problems that confront them etc.

> Insofar as all our thinking…is under the limitation of a horizon, then, first of all, the suggestions and motivations that arise from the situation are given a twist by the limited mentality of that horizon. People will see what they want to see, what can fit within their horizon, and they will omit the rest, or at least they will omit the significance of the rest. Insofar as thinking, reflecting, deciding and policy-making are under the limitation of a horizon, there is the recurrence of overemphasis and oversight in the consequent situation. The succession of situations, then, will reveal the cumulative effects of the limitations of this horizon…The result is that the situation progressively deteriorates.
>
> (Lonergan [1957] 2001c: 304)

The products of each group, whether activities, language, social structures or practices, policies, mores, art, music etc., will express their horizon. The history of these products will be a history of both the achievement of a society and a culture as well as a history of their distortion and (mis)use by the dominant group. Economic, social and political structures and their associated policies and practices will be some mix of achievement and distortion.

Within the horizon of everyday common-sense living, conflicts come and go, dormant at times but fiercely contested at others. Sometimes a group can maintain its position by force or the exertion of dominating power. So a landlord can lobby for legislation that supports their position or impose new policies and practices on tenants. Sometimes a group can, in their own interests, accommodate another group, so a landlord forms coalitions with the housing construction industry or a political party. Sometimes a new group forms in opposition to the dominant group and can eventually replace that group over decades or centuries, so social housing, as a response to housing poverty and the failure of the private rental market, eventually becomes an accepted housing tenure, albeit in a limited way. But resolutions such as these are short term and cyclical as one dominant group replaces another, as one blind spot replaces another, as one management style replaces another, as one policy priority replaces another.

The diversity of horizons in everyday common-sense living, however, is an ongoing source of conflicts between groups, and we continually face the problem of working through these conflicts as we work out how to live together.

There is, however, a further problem that allows conflicts between groups to continue even as the source of the conflict changes: everyday living demands immediate solutions; it ignores the long-term consequences of these solutions; it is incapable of doing the hard work of understanding the problem and its complexities. By operating within the horizon of everyday common-sense living, our horizon is restricted:

> The simple fact about human life...is that the problems encountered most regularly throughout human life demand a general level of developed capacities and skills in excess of that which is commonly operative. This fact is true not only of aggregates of persons, but also of the course of any one person's life...
>
> (Melchin 1999: 233)10

The scope of understanding within the horizon of everyday common-sense living is limited. It is incapable of analysing itself and recognising that it is a specialised area of knowledge, one that is concerned with particular situations now (Lonergan [1957] 1992: 251).

Housing researchers operating within the horizon of everyday common-sense living face two particular problems. First, their understanding of housing depends upon their particular horizon of everyday common-sense living. This horizon is one among many. It directs attention to some events rather than others. It regards some events as more important than others. It incorporates the interests of one group rather than another. Rather than contributing to a resolution of conflicts, housing researchers are themselves involved on one or other side of the conflict. Second, in common with all groups, housing researchers are unable to distinguish between housing and its use by different groups for their self-aggrandisement i.e. how it is used to preserve and expand their interests. It is this inability – indeed, resistance – to distinguish housing from its uses that prolongs conflicts as dominant groups shift over long periods of time.

Although housing researchers may in some way recognise that there is a problem of ongoing and unresolved conflicts, their limited horizons prevent them from finding ways in which to resolve them. If we are to distinguish the practices that are constitutive of housing from the associated practices of the dominant group, we need to turn to a different way of understanding the situation. In short, we need a higher viewpoint on these conflicting horizons, one whose ultimate purpose will provide a better insight into what is happening by distinguishing various elements such as housing and its associated uses. This higher viewpoint is the horizon of theory.[11]

Horizons of theory

If housing researchers are to grasp more fully the situation that confronts them in their everyday living and if they are to intervene effectively in that situation, they need a higher viewpoint; they need a viewpoint which reaches beyond the horizons of particular economic, social, political, cultural and religious groups and which reaches beyond the restricted horizon that elevates immediate solutions above long-term solutions that meet the complexity of the situation.

This higher viewpoint is brought about by the pursuit of understanding through asking and answering questions. In this pursuit our orientation shifts. We are no longer satisfied with descriptions that get us through our everyday living. How things relate to us becomes secondary; it becomes a platform for further and prolonged questioning. Our orientation here is towards adequate understanding. It is an orientation in which we prolong the questioning process until we are satisfied that we understand what it is that we are investigating. We make distinctions between one thing and another, not on the basis of how they appear to us, but rather on the basis of their relationship to other things. Our orientation is towards *an understanding that relates things to one another*, so, for example, we distinguish rent-setting from the attributes and characteristics of those who pay or receive rents, from how the level of rent is determined, from its role in achieving affordability, from its role in distributing benefits equitably among tenants or households, from its role in relation to the production of other goods. We can then proceed to relate rent-setting to a finance system, to social housing, to housing affordability, to an economy etc. Within this orientation, we can understand the relationships between rent-setting and other aspects of a society. Thus, we can move towards solving more complex problems by providing long-term solutions.

Through the pursuit of questions we reach an understanding of what housing is, a grasp of what constitutes housing, what distinguishes it from other things. It is a grasp which goes beyond the various understandings and perspectives of those operating within the horizon of everyday common-sense living.

This orientation towards pursuing and answering questions is a shift in horizons, a shift from the world of everyday common-sense living to the world of theory. But such a shift is not easy. This shift in fundamental horizon demands a new way of thinking, a new way of asking questions, a new way of operating. The primary difficulty is the long-held and stubborn assumptions that inhibit asking questions and prevent new insights. These are the assumptions of a culture, both the culture of the society in which the scientist operates and also the culture that is particular to each science. The world of theory challenges (and overturns) long-held assumptions about how things work.

In part, the history of modern science can be understood as a history of discovery due to more sophisticated measuring techniques and devices, but it is also a history of discovery as long-held assumptions and presuppositions are slowly overturned and revealed in a revolutionary shift in thinking. Herbert Butterfield (1957), for instance, documented the revolution that overturned the long-held Aristotelian physics and biology as well as Ptolemaic astronomy. Thomas Kuhn (1970) documented the scientific revolutions brought about by Copernicus, Lavoisier, Galileo, Newton and Einstein. As Lonergan notes, each revolution in thinking brings with it a crisis and a shift in scientific horizon:

> The scientific horizon recedes, expands, where there occurs a crisis in existing methods, procedures, theories, assumptions which are seen to fail. They cannot handle known results, known observations or data, known conclusions. The crisis arises from a fundamental conflict between basic assumptions or methods or presuppositions and, on the other hand, something that within that order of investigation has to be accepted, something of the order of fact or inevitable conclusions. Upon this crisis there follows a radical revision of basic concepts, postulates, axioms, methods, and a consequent new mathematical or scientific structure.
>
> (Lonergan [1959] 1993: 92–93).

While the achievements of the natural sciences have revolutionised our understanding and our living, the achievements of the human sciences have been much more limited. While the natural sciences have been able to resolve conflicts (at least provisionally) and move on from crises, the human sciences have been less able to do so. In the human sciences, it has been much more difficult to reveal the assumptions and presuppositions of researchers.

Within the social sciences, conflicts abound as different understandings and theories of society develop: some in overt support of different dominant groups (market economics, capitalism, militarism, sexism, central government etc.); some in overt support of oppressed groups (Marxism, feminism, co-operative economics, democracy etc.); and others supporting understandings of society such as functionalist, behaviourist, social constructionist and voluntarist. Different philosophies develop with differing positions on the status of our knowledge, on what we can aspire to: empiricist, logical positivist, idealist, critical realist, naive realist, phenomenologist, existentialist, scholastic. Within these differing philosophical positions is the conflict between the goal of objectivity and the inability to exclude subjectivity. Different disciplines develop different perspectives on persons in society and utilise different methods and heuristics. Theoretical achievements are not once and for all. One line of understanding develops and consolidates its position, only to be supplanted by another. Just as our everyday living continually changes, so too does our understanding of that living and of the things we seek to understand. Any achievement is provisional because it is always subject to further development. Throughout all these differences, there is the ongoing tension between, on the one hand, theory with its demand for a more adequate understanding and, on the other hand, everyday living with its demand for immediate practical solutions to current problems.

If we, as researchers, are to reach some integral position on these different, ever-changing and provisional theories, then we need a higher viewpoint, one from which these conflicts within the social sciences can be resolved.

But what would such a higher viewpoint be?

Horizon of the subject as subject

Above I asked the question: what is theory? In answering that question, I turned to – and proposed that housing researchers turn to – the data within themselves and pay attention to what is going on when we are theorising. This viewpoint on theory is developed within the horizon of the subject as subject.[12]

Within the horizon of everyday common-sense living, we focus on the immediate practicalities of everyday living, on what we intend to do, whether solving everyday problems or creating something new. Within the horizon of theory, we focus on understanding, on what we intend to research. Within the horizon of subject as subject, however, we focus on the intending of the subject by which such doing and understanding come to fruition. The horizon of the subject as subject is not concerned with the values of the researcher, nor their identity, nor their knowledge or their heuristics, nor with housing researchers articulating the values and beliefs they bring to the research. These are the products of their activities, the products of the society and culture within which they live. Rather, the horizon of the subject as subject is concerned with what is prior to these values, identity and knowledge; it is concerned with the dynamics of the subject that makes these products possible. The horizon of subject as subject marks a shift away from the products of everyday living (technology, economy, society, politics, culture and religion) and the products of theory (the expression of a theory of housing etc.). It shifts away from these human expressions towards a heightened awareness of the subject in action. The subject as subject is the

> prior ontic[13] reality in which one is going to find the norms and invariants that are common to all horizons, that recur in all subjects, the reality that provides the real norms on the basis of which one can select the true horizon…It lies in the subject that does the talking, and not in the subject he talks about…The philosophers will differ in their accounts of knowledge, in what they say of themselves in their account of the subject as object, but that will not prevent them from being the same type of being, where the word 'being' is used not ontologically but ontically, in the sense of the reality you know in some fashion, in some, within quotes, sense of the word 'know' prior to conceiving it and affirming it.
>
> (Lonergan [1957] 2001c: 314–315)

For example, when I do research, there is present to me not only my research (what I am researching) and my researching (my activities) but also myself as researcher (the I who am researching). Similarly when I see marks on a page, there is present to me the seen (the marks on a page), the act of seeing and the seer (the I who am seeing). All three – the object, the activity and the subject – occur together simultaneously. The horizon of subject as subject is the I who am researching or seeing, it is the concretely operating subject.

We can come to an understanding of ourselves by researching others or ourselves. Then we come to an understanding of the subject as object. An understanding of the subject as subject, however, is reached not through research as commonly understood, but through a personal process of heightening our awareness of ourselves as subject as we undertake activities – it is a matter of 'catching ourselves in the act'. So, for example, in the process of doing research we heighten our awareness of ourselves as subject in this doing. By heightening our awareness of ourselves as subject, we can go on to explicate the dynamics or desires operative within the subject, dynamics which reveal the subject as subject. Such dynamics would include the desire to understand which underpins all our learning, researching, discovering and knowing; the dynamic of creating something worthwhile which underpins the expression of ourselves in the development of symbol and language, arts, technology, economy, society and culture; the dynamic of sociality which underpins our living together and which binds together families, communities and societies; the dynamic of being in love which underpins our identity and the world we live in.

By heightening our awareness of these operative dynamics we are asking and answering the question 'who am I?' And, in so doing, we are asking and answering the questions: 'who am I as a housing researcher?' and 'who are we as collaborating housing researchers?'

Dialectic as a method for searching out the best of the past

Above I noted three issues in relation to evaluation and critique: the need for a framework that will inform the decisions and actions of individuals and groups (and get us beyond the distorted practices of dominating power and the self-aggrandisement of individuals and groups); the need for a framework that will inform the decisions and actions of housing researchers (and get beyond the conflicting understandings between individuals and groups); and the need for a framework for those researchers undertaking evaluation, comparison and critique, whether of individuals and groups, housing researchers, or housing researchers undertaking evaluation, comparison and critique.

If we retrace our steps, we discover that most of our decisions presuppose some prior decisions. I have referred to these prior decisions as a framework or horizon within which we make our current decisions. This horizon intrudes upon and orients our current decisions. So, for housing researchers, this horizon not only intrudes upon decisions about what to investigate, decisions about method and decisions about heuristic, it also intrudes upon their findings.

I have proposed a view of theory which is grounded in the horizon of the subject as subject. In addition, I noted that the explanandum of theory is a value, something that is worthwhile. Further, that a theory of housing locates housing within a hierarchy of values.

I also distinguished three horizons: the horizons of everyday common-sense living, the horizons of theory and the horizon of the subject as subject. Each of these operates in the background, orienting our lives and decisions. But the horizons of everyday common-sense living and the horizons of theory are also a source of fundamental conflicts between individuals, between groups, between housing researchers.

The conflicts that ensue from our horizon are very difficult to bring to light because that horizon largely remains hidden. The role of Dialectic is to deal with these fundamental conflicts, to bring to light the horizon within which subsequent decisions are made and to provide a method that objectifies the operative horizon and promotes personal transformation (Lonergan [1972] 1990: 235).

What I am proposing here it that the objectification of the third horizon – the subject as subject – provides a critical framework for (i) individuals and groups operating within the horizon of everyday common-sense living, (ii) housing researchers operating within the horizon of theory and (iii) housing researchers undertaking evaluation, comparison and critique of individuals and groups, of housing researchers and of themselves.

Further, the goal of Dialectic is to resolve these fundamental conflicts. Such resolution is not simply a matter of developing a new position and refuting others: 'If Descartes has imposed upon subsequent philosophers a requirement of rigorous method, Hegel has obliged them not only to account for their own views but also to explain the existence of contrary convictions and opinions' (Lonergan [1957] 1992: 553).[14] Morelli (1995: 379), in reflecting upon this comment, notes that the

> Hegelian requirement is a demand for a radically heightened philosophic reflexivity, for philosophers henceforth to do more than produce positions, acknowledge other opposed or different positions, and then attempt to refute those other positions without regard either for their genesis or for the methodological limitations of such a controversialist mode of argumentation.

No longer can housing researchers just develop and argue for a position and refute alternative positions. What we need, then, is a method which accounts for opposing positions by integrating them within a higher viewpoint. The method by which we do this, Dialectic, is not simple however.

Dialectic proceeds by way of two movements, a movement of appreciation which recognises the achievements of each position and a movement of critique which recognises their limitations. It is through this process of Dialectic that the best of the past is discovered, rediscovered and retained and that progress is achieved and consolidated.

The following expounds on this proposal. It considers the frameworks of housing researchers, housing practitioners and those researchers evaluating and critiquing researchers and practitioners.

Dialectic and housing researchers

Housing researchers deliberate about and make decisions to undertake research. They then pursue answers to questions. Underpinning this pursuit and underpinning all research is a dynamic within the subject, the dynamic of the desire to understand.

For the researcher asking and answering questions, becoming self-luminous about the dynamic of this desire to understand (which produces knowledge) is a long process of being attentive to oneself, to one's own internal processes and experiences (the activities of seeing, hearing, feeling etc.), grasping the relationship between certain activities and research, personally verifying that these are the significant, relevant and essential elements that constitute research. To reach the point of personally verifying this understanding of the dynamic of the desire to understand, a researcher makes three affirmations. First, an affirmation that distinguishes the occurrence of various activities within themselves such as seeing, hearing, feeling, imagining, considering, formulating, questioning, understanding, affirming etc. Second, an affirmation that the occurrence of some of these activities are related to one another: that questions are questions about events, that insights are insights into events, that affirmations are affirmations of insights; that the occurrence of selected events within themselves (experiences, questions, insights and affirmations) are the significant, relevant and essential elements that together constitute knowledge; and that other events that occur within themselves are not significant and not relevant to research. Thirdly, an affirmation that this is the best account of research because they have attended to their experience of these internal activities, they have asked all the relevant questions about their experience and have reached an understanding that makes sense of their experience.[15]

While these activities are internal to each housing researcher and their verification is through a personal experiment, they can be objectified, made public and communicated in words. As a meaningful human expression, these words are interpreted (understood), critically reviewed, revised, affirmed and acted upon by others. Its successful acceptance by others depends not only on whether the dynamic of the subject as subject is adequately grasped but also on an apt choice of expressions within a particular cultural context. As a human expression it is both an expression of something and a social construction dependent upon the preceding development of language and the individual's level of skill in using this language. Thus, any explication of this dynamic is limited. It has its origin and source in the subject as a subject. While the explication of this dynamic is expressed in theoretical terms, it is grounded in the subject as subject.

An investigation of this dynamic reveals that the desire to understand is constituted by a set of related operations or activities (Lonergan 1967, [1957] 1992: Chapter 1, [1972] 1990). It is this dynamic of the desire to understand that gives rise to the horizon of theory in which things are understood not in terms of how they relate to us but in terms of how they relate to one another.

This dynamic is common to all methods within both the natural and the social sciences (Lonergan 1967, [1957] 1992, [1972] 1990: Chapter 1). It underpins the development of all the different skills, techniques and methods of our researching that lead us to new discoveries. It underpins the range of methods in the many and varied sciences, that adapts old methods and develops new ones for undertaking research.

This dynamic of the desire to understand continually moves our understanding beyond where we are now to some better understanding. It is not content with just experiencing the world; it wants to understand it. It is not content with just understanding what is possibly happening; it wants to understand what is actually happening. It is not content with partial understanding; it wants to understand all that can be understood. It is not content with ambiguities or contradictions or unanswered questions or conflicting views; it wants to resolve them. It is not content with understanding this and understanding that; it wants to understand this in relation to that (for example, housing in relation to the economy and society).

The desire to understand seeks an integral view in which all aspects of housing can be understood in relation to one another. But such a view is not reached immediately or even in a short time. It is reached after an extended period of asking and answering questions about data, gaining insights into the data, correcting previous oversights, relating one insight to another and one set of insights to other sets of insights, and affirming not just one insight but sets of related insights. So an integral view is something yet to be attained, and progress towards it is a step-by-step process which builds upon previous achievements and undoes the deceptions, unwarranted presuppositions and oversights of the past. The pursuit of a question brings with it a higher viewpoint on the data, and progress occurs in accordance with the dynamic of the subject.

So, for example, different methods emerge as housing researchers seek to understand the complexity of housing. Questions then arise as to: which method is used in which circumstance; the type of question each method is seeking to answer; how these methods relate to one another; whether an integrated view of all these methods can be reached. By grasping the dynamic of the desire to understand we can distinguish different types of questions and anticipate the type of answers that each intends. In this way, we can appreciate the primary question that each method is oriented around but also criticise them for shifting from question to question. In doing so, we can relate these apparently disparate methods to one another around their primary question.

I have noted how housing researchers undertake particular types of research forming different genres. To varying extents they regard their type of research as pre-eminent. So, for instance, housing research is caught up in ongoing methodological and philosophical disputes such as those between positivists who give pre-eminence to quantitative data; social constructionists who give pre-eminence to the constructed perspectives of participants and to the constructions of theorists; and critical realists who give pre-eminence to the underlying causal mechanisms of necessary and contingent relations. They each presuppose some position on housing research. They each presuppose a decision by housing researchers to act in accord with this position and to do their research in a particular way.

Above, I referred to the work of Butterfield (1957) and Kuhn (1970) who showed how the inability of a current theory to explain new evidence brought about a crisis within a science which was only resolved by a new paradigm which integrated both the old and the new.

How, then, are we to reach some integrated understanding that not only accounts for the view that these are different genres of research, but also 'explains the existence of contrary convictions and opinions'?

An understanding of the range of dynamics operative within the subject as subject provides a heuristic for such an explanation. On the one hand, the dynamic of the desire to understand unfolds on three levels – experience, understanding and affirmation – within the context of a fourth level, the decision to undertake some research project that is worthwhile. So, while this dynamic recognises the significance of experience to research, it does not reduce research to experience (to what can be seen, heard, touched, felt or imagined) as empiricists do; while it recognises the significance of creativity and constructive insights to research, it does not reduce research to these insights, as idealists and social constructionists do; while it recognises the significance of affirmations to research, it does not reduce research to affirmations, as fideists do. Rather, within a larger framework, research is constituted by integrating all three – experiencing, understanding and affirmation – with each level concerned with a particular question.

Further, the dynamic of the desire to understand can reflect upon the past or it can look to the future. We can therefore distinguish empirical, theoretical and historical questions which reflect upon the past from practical, strategic and policy questions which look to the future. We can also distinguish between an evaluation/ critique of past frameworks for research (and the struggle to integrate diverse philosophical and ontological approaches, disciplines and methods) and the transformation of future research as these past frameworks are integrated into a higher viewpoint. At this time, this higher integrated viewpoint is Functional Collaboration.

On the other hand, an explanation of contrary positions regards not just the dynamic of the desire to understand but its interference by other dynamics operative within the subject: the dynamics of creating something worthwhile, of sociality and of being-in-love. To these we could add the dynamics of the psyche, of the subject as a living organism, of the subject as chemical and physical processes.[16]

These operative dynamics constitute different aspects of the subject. Each dynamic brings about something of value. Each operates within a context, with each dynamic related to others, each playing a role in the unfolding of the subject. Thus, if we want to understand the dynamics that constitute a researcher, a social agent, a subject, then we can develop an ordered set of theories in which lower level dynamics have their own autonomy and provide the conditions for the emergence of a higher level dynamic, and the higher level dynamics order the particularities of lower level dynamics. As noted previously in both Chapter 5 with reference to a theory of housing and above in this chapter with reference to a theory of theory, this ordered set of theories is a particular type of hierarchy, one not based on dominance but on mutual interaction.

An integral view of the subject is one where we can reach an understanding not just of each dynamic but also of their relations to one another. An integral view of the subject is a hierarchy of dynamics. So, the dynamic of the desire to understand has a role in the dynamic of creating something worthwhile, the dynamic of creating something worthwhile has a role in the dynamic of sociality, the dynamic of sociality has a role in being-in-love. But being-in-love orders (or systematises) the particularity of the elements that constitute the dynamic of sociality, the dynamic of sociality orders the particularity of the elements that constitute the dynamic of creating something worthwhile, the dynamic of creating something worthwhile orders the particularity of the elements that constitute the dynamic of the desire to understand, the dynamic of the desire to understand orders the particularity of the elements that constitute the dynamic of the psyche etc. Each person, the subject as subject, is seeking to integrate these dynamics.

As a consequence, it is the deliberation and decision of the researcher to answer one question (as creating something worthwhile) rather than another that precedes their pursuit of answers to the question. This decision has its grounds in an affirmation or discovery that understanding creates something worthwhile. The decision to operate in accord with the dynamic of the desire to understand gives effect to this discovery and, once a decision is made to answer a question, the dynamic of the desire to understand become operative. Yet, the researcher can decide to change their question; they can call a halt to research and shift into another dynamic of the subject, such as the demand for an immediate practical response.

This capacity to shift from one dynamic to another introduces the possibility of conflicts as the subject seeks to integrate the range of dynamics within a single entity, the subject. Where the set of related operations or activities for one dynamic is not distinguished from another, it introduces the possibility that one dynamic can interfere with another. So, a researcher may select a particular heuristic that will guide their research (a technological heuristic or an economic heuristic or a political heuristic etc.) without recognising its limitations, or may cut short their pursuit of answers to a set of questions because they sense that the answers will place them at odds with colleagues, because they are contrary to long-held views, because they threaten his or her position, because they are at odds with the views of the dominant group etc.[17]

A decision to operate in accord with the demands of the subject as subject requires a radical transformation of the researcher. A researcher can live and develop quite remarkably within the horizon of everyday common-sense living. But the shift from this horizon to the horizon of theory and the shift from the horizon of theory to the horizon of subject as subject are radical transformations, indeed, displacements from one world into another.[18] This decision marks a basic conflict – the conflict between operating in accord with one dynamic and its interference by other dynamics. For the researcher, this conflict emerges as the conflict between the dynamic of the desire to understand and the demands of other dynamics – the dynamic of being-in-love, the dynamic of sociality, the dynamic of creating something worthwhile, the dynamic of biological sustenance etc.

The decision of a housing researcher to operate in accord with the dynamic of the desire to understand and to shift to the horizon of theory is not an easy one, nor is it easy to put into practice. It demands a high level of being self-luminous about what we are doing within the dynamic of the desire to understand, and how other dynamics can interfere with this dynamic. Moreover, our living is constituted by a range of dynamics and we move from one to the other rapidly. A researcher continually moves from one horizon to another. At any time, we need to be self-luminous, self-conscious of the horizon within which we are operating.

The difficulty in the displacement into the world of theory is illustrated by the acknowledgement of the need for theory, the ongoing debates about theory, the lack of clarity about theory and the difficulty researchers have in distinguishing the world of theory from the world of everyday common-sense living (amid the obfuscation of jargon).

Again, the difficulty in the displacement into the world of the subject as subject is illustrated in the slow process of understanding the dynamics of the human person: the dynamic of understanding, of creating something worthwhile, of sociality and of being-in-love. The displacement into the world of the subject as subject demands a shift in attention from the products of human expression to the I who precedes these products, who is doing the expressing.[19]

While these radical transformations are the achievement of each person, they are achievements which are cultivated by and sustained by the group or scientific community (Kuhn 1970). It is the community that extends the invitation and the challenge to explore new horizons. It is within the community that the fruits of these radical transformations are realised.

Just as the dynamic of the desire to understand puts demands on researchers, so too does the dynamic of creating something worthwhile. It is a demand that continually moves researchers beyond where they are now to something better. It is not content with current achievements. It is not satisfied with what pleases or feels good; it is not satisfied with what is apparently good or what family, friends, communities, societies or cultures uphold as good. It heads beyond this to what is worthwhile. But as I noted above, a researcher pursuing an answer to a question creates something worthwhile, but what is more worthwhile is the decision to act upon the basis of this discovery.

For housing researchers (as researchers) a fundamental decision is whether they will act in accordance with the dynamic of the desire to understand, the dynamic operative within them. It is a dynamic that calls for the ongoing pursuit of answers to questions, the resolution of conflicting answers and the development of an integral view. Insofar as they act in accordance with the dynamic of the desire to understand, they are competent researchers or, to use the existentialists' term, they are acting authentically. As a consequence, their results are sound. Insofar as they allow other dynamics of the subject to interfere with the dynamic of the desire to understand, they are not acting as competent researchers and their results will be unsound. Insofar as their decisions are grounded in their discoveries, researchers are competent in creating something worthwhile, viz. an understanding of housing.

In this way, the horizon of subject as subject becomes the criteria for evaluating the horizons of housing researchers.

Dialectic and housing practitioners

Housing researchers using the methods of evaluation and socio-historical critique are seeking something better than the current state of affairs. These evaluations/ critiques encompass both the outcomes that have been created and the practices whereby they were created.

The problem confronting housing researchers using the methods of evaluation/ critique is whether or not their frameworks are arbitrary. Generally, they juxtapose their framework against that of those who created this state of affairs. They presuppose that their framework is superior in some way: it may purport to be a neutral framework or it may support the interests of a particular group (whether a dominant group or discriminated group). How, then, can we come up with an alternative framework which is not arbitrary?

Previously, I distinguished between what housing is and the purpose or role it plays. Housing and each of its purposes is something of value, something worthwhile. So, while a theory of housing is the set of related elements that constitute it, a complete theory of housing is an ordered hierarchical set of sets of related elements that incorporate the whole range of its purposes and orders its characteristics. This complete theory operates as a heuristic for understanding actual operating housing systems at any particular time and place.

In Chapter 6 on History, I outlined an understanding of history as grasping the vector that motivates the shift from one actual operating housing system to another. Further, I noted that the vector that motivates the shift is one of the values within the set that constitutes a complete theory of housing and that this vector itself may change over time. This vector not only re-orders or re-systematises the particularity of the elements that constitute this vector but it also re-orders the particularity of the elements that constitute the lower values and also provides a set of new conditions for the achievement of higher values.

The question for Dialectic is whether this vector of change is a development or a distortion of housing as a whole, whether it promotes the totality of housing purposes. The vector of change brings about a re-ordering of the particularity of the elements that constitute housing. What is subject to evaluation and critique here is not so much the value *per se* but its realisation in the particularity of the elements that constitute it. Insofar as this particularity contributes to the whole, to the totality of housing purposes, it is a development or improvement and is worthwhile. Insofar as it undermines the whole by giving priority to the realisation of one value, it is not a development.

Within the horizon of theory, we can understand the exercise of dominating power and aggrandisement of self or group as linked with giving priority or exclusiveness to one value within the hierarchy of values rather than to the achievement of the whole. For this individual or group, the primary purpose or role of other values with the hierarchy of values is the achievement of one particular value. It is the priority afforded the achievement of this one value that creates an ideology. In seeking its own self-aggrandisement and interests, a group will promote one of the values within the hierarchy of values (within the theory of housing). In this way, we can link the interests of a particular group with the priority accorded to the realisation of one of these values in this ordered set of values. In doing so, they seek to determine the particularities of housing in accord with their own interests in this value. Conflict ensues as different groups align with different values in this ordered set and seek to determine the particularities of housing in accord with their own interests. Each value has a legitimate role within the whole. It is this very legitimacy that makes this ideology so difficult to grasp as the exercise of dominating power and aggrandisement of an individual or group. It is this legitimacy that makes the ideology so stubborn to shift and to reverse. The difficulty lies in grasping its ambiguity. The value promoted by the dominant group is legitimate but within a large whole. Yet, it is promoted to the detriment of this larger whole and it is only by grasping this larger whole that we can come to some understanding that its particularities are detrimental to the progress of the whole.

This view of ideology begins to make sense of ongoing conflicts. For example, it begins to make sense of the ongoing conflict in Australia between social housing tenants seeking housing affordability and social housing providers seeking financial viability. In the theory of rent-setting outlined in Chapter 5 (and illustrated in Figure 5.2), I noted that financial viability and housing affordability are values within a hierarchy of values in which rent-setting plays a role. Further, that the set of related elements that constitute rent (specified sets of data, the specific operations carried out upon these specified sets of data, and the application of the results of these operations to a specific tenancy) are particularised in such a way that both these values (along with others) will be achieved. An evaluation of the history of rent-setting in Australia reveals that the particularisation of the elements of rent-setting has given priority to financial viability (in the interests of governments and social housing providers). On the other hand, it also reveals that housing tenants have given priority to housing affordability without due recognition that rent-setting must also achieve the financial viability of social housing. While financial viability is a legitimate value to be achieved, it becomes an ideology when the particularities of rent-setting are ordered in such a way that financial viability is given priority over the whole set of purposes for rent-setting. In this way it serves the interests of or aggrandises certain individuals or groups.

The issue here is not the value *per se* but its realisation in a particular set of related elements within a total set of sets of related elements. Evaluation seeks to determine whether and to what extent the realisation of housing in history has given priority to a lower value over a higher value in the ordered hierarchy of values or has given priority to a higher value but in the course of doing so has ignored the exigencies of the lower value.

Within the horizon of theory where we distinguish between what housing is and its role in constituting other values, we can understand the particular role which housing plays in the achievement of some other value. History reveals the vector that is the motivation for shifting from one form of housing to another form of housing. It is this vector which Dialectic is seeking to evaluate – does this vector promote progress, a development which integrates the whole range of values in which housing plays a role? The dynamics of the subject as subject place a demand upon the particularisation of the related elements that constitute housing, viz. that what is created is worthwhile, better than the previous particularisation. It will only be better where the change brings about a development which integrates the whole range of values in which housing plays a role. It is the intelligent and responsible course of action, one which is the best that we can come up with at the time.

Housing practitioners whether governments, organisations or individuals make their current housing decisions within a horizon of everyday common-sense living. It is the operative framework within which they deliberate about and decide what to do. Embedded within this horizon are presuppositions that maintain, support and promote particular values or give priority to some values over others. Dialectic, which is concerned with decisions insofar as they point to a set of prior decisions, seeks to bring this horizon into the light and to evaluate it. Housing researchers operating within the horizon of theory use a theory of housing (the complete ordered hierarchical set of sets of related elements that incorporate the whole range of its purposes and orders its characteristics) to evaluate the extent to which the vector that informs the decisions of housing practitioners will promote progress or deterioration in housing.

Dialectic and the grounds for critique and evaluation

Both housing practitioners and housing researchers implicitly operate within the horizon of the subject as subject. A decision to take a stand within one or other group, one or other discipline, one or other philosophy, one or other understanding of theory, one or other developing view of living and either the horizon of everyday common-sense living or the horizon of theory has implications for our understanding of housing, implications for what new policies and practices we might implement to make housing better. Our problem is not simply one of philosophy and philosophical positions. Rather, our problem is a personal one. It is a problem of deciding how I am going to concretely operate as a researcher.

For a researcher, the fundamental conflict inherent within social science is a conflict between who I am as a person in society and who I think I am:

> The conflict lies…between, on the one hand, what man is and is to be and, on the other hand, what man thinks he is and is to be. The conflict is not between two schools of thought but between human reality and human thought about human reality. It is human thought about human reality that produces the technical, social, and cultural situation. It is the limitations in human thought about human reality that produce the evils in the technical, social, and cultural situation. And it is the revolt of human reality (as distinct from human thought about human reality) against those evils in the situation that provides the lever for correcting the defects in the human thinking about man.
>
> (Lonergan [1957] 2001c: 308)

Any new discovery not only has implications for what is being investigated, it also has personal implications for the housing researcher. It challenges their understanding of themselves, of who they are. Consequently, it invites them to live in a new way that is concomitant with their new understanding of themselves. It is an invitation to further development. But it is also an invitation that can be refused and in that refusal there lies a closing-off of further questions and further answers.

Our horizon is an achievement of past decisions whereby we have learned to organise our world, whereby we have learned to operate more or less successfully within the larger world in which we live, whereby we have spontaneously, even unconsciously, made ourselves who we are. This spontaneous development can become more or less deliberate as we consider who we are. We can make decisions as to what interests us, what is important to us, what we care about. Insofar as we do this, our horizon shifts. A housing researcher can make a decision as to the horizon within which they will operate: the horizon of everyday common-sense living, the horizon of theory or the horizon of the subject as subject.

Moreover, once we advert to the connection between our researching and our horizon, we can raise a question for deliberation: 'whether we want to make a change or not?' Or, to put it another way, 'whether we want to take full responsibility for who we are or not, who we will become or not?', 'whether we will become active agents, self-constituting subjects?' (Lonergan [1957] 2001b: 293–294).

Change in any significant way is not easy. It demands a reorganisation of the subject, a re-orientation in living, thinking, feeling; a change in what we do and how we do it; a change in the groups with which we share our researching.

> Against such reorganization of the patterns of the subject, there comes into play all the conservative forces that give our lives their continuity and their coherence. The subject's fundamental anxiety, his deepest dread, is the collapse of himself and his world. Tampering with the organization of himself, reorganizing himself, gives rise to such a dread.
>
> (Lonergan [1959] 1993: 90)

What is required of housing researchers is that they search out and retain the best of the past (that which is the product of the pursuit of answers to questions) and discard what is shown to be inadequate, irrelevant and destructive (that which is the product of the interference of the dynamic of the desire to understand with other dynamics). There are, however, no *a priori* answers as to what is best. In the great experiment of history we have to work this out ourselves. It is by 'retracing our steps', by reflecting upon, understanding and evaluating our history, that we learn what promotes progress. Progress is not a logical process nor is it fixed for all time. It is ever on the move as the future builds upon the discoveries of the past.

While the horizon of the subject as subject does not provide *a priori* answers as to what is best, it does provide a heuristic of progress as that which is in accord with the dynamic of the subject. At any time, the future is forged from the best we can come up with about the past. So, to promote progress we need to live in view of the best of the past.

History reveals the flow of changes in this past and what drives this flow of change. However, History does not distinguish between what is progress or development and what is decline or deterioration in this past. As a pointer to future directions, housing researchers need to work this out. To do so, they need a method that investigates whether they have operated in accord with the dynamics of the subject as subject.

The thematisation of the subject as subject provides us with criteria for distinguishing understanding from not understanding, creating something worthwhile from something that is not worthwhile etc. In themselves the dynamics of understanding and creating something worthwhile are open; they do not point to any particular understanding or product as progress.[20]

Housing researchers operating within the functional specialty Dialectic critique and evaluate the historical movements within everyday living and within housing research. They indicate how the materials produced by Researchers, Interpreters and Historians might be viewed if they proceeded from a radical transformation of their horizon. In doing so, they reveal their own grounds for progress. This is essential to the ongoing process of coming up with the best of the past. So, the results produced by one housing researcher within the functional specialty Dialectic become materials to be worked on by other researchers within the functional specialty Dialectic.[21]

Some illustrations of Dialectic

Illustrations of Dialectic are hard to come by. This book itself is a minimalist version of Dialectic as it proposes a way in which to integrate a range of disparate and often conflicting methods within the social sciences. It seeks to affirm the core elements of these methods through a phenomenology of research while (implicitly) critiquing other elements. Within the larger framework for collaborative creativity, Functional Collaboration, it seeks to define and relate these core elements of differing methods. In my discussion of 'Dialectic and housing practitioners' I drew upon the indicative theory of rent-setting outlined in Chapter 5 to illustrate the differing horizons of housing practitioners, indicating how these horizons picked up different aspects of housing and how they could be integrated into a larger whole. In Chapter 5 on the complete context of housing, I sought to develop a more adequate heuristic for a theory of housing, one which integrated the findings of various disciplines: technology, economics, politics, cultural studies, personal identity. This illustration again seeks to affirm the core elements of each of these disciplines (the best of the past) and implicitly critiques other elements. It then integrates them into a more adequate theory of housing.

Gibson Winter (1966) in his *Elements for a Social Ethic: Scientific Perspectives on Social Process* provides another illustration of dealing with conflicts within social science by shifting to the horizon of subject as subject (in this instance, the dynamic of sociality). Winter assembles four conflicting social theories of the 1950s and early 1960s: behaviourist, functionalist (Talcott Parsons), voluntarist (C. Wright Mills) and intentionalist. He begins by exploring the core problem of social reality, our intersubjectivity. This was first formulated as the 'I' and 'me' problem by George Herbert Mead (1938). He then draws on the work of Alfred Schutz (1972) and explicates what we are doing as we relate to one another. It is within this framework that he analyses these four conflicting social theories, goes to the root of their difference and recognises their limitations but also their strengths. His explication of the process or method of sociality allows him to relate and ultimately integrate these four conflicting social theories within a higher viewpoint.[22]

At a more sophisticated level, Lonergan also offers a variety of illustrations of Dialectic: his integration of various forms of philosophy (Hume's empiricism, Kant's critical idealism, Hegel's absolute idealism) (Lonergan [1957] 1992); the integration of classical, statistical, genetic and dialectic methods (Lonergan [1957] 1992); Freud's unconscious and censorship (Lonergan [1957] 1992: Chapter 6, Section 2.7); and economic theory (Lonergan [1942–1944] 1998, [1983] 1999; see also McNelis 2010).

Implications for housing research

Dialectic is a method whereby we 'retrace our steps', we encounter the people who have made the past and we search out what is the best of that past. In doing so, it deals with the conflicts in that past and seeks to resolve them by integrating what is worthwhile into a larger perspective.

I began this chapter by noting how a decision is implemented through a further series of decisions. So, when a housing researcher makes a decision to investigate a particular area of interest or to pursue an answer to a particular question, this leads to a whole series of decisions as to how to proceed. I identified four genres of housing research that are concerned with the decisions made by housing practitioners and housing researchers: evaluation studies, socio-historical critiques, comparative studies and methodological critiques. These genres presuppose that housing practitioners and housing researchers could produce something better than what they have already done. They presuppose some framework within which housing practitioners and housing researchers could operate but, to the extent of their critique, they do not. They juxtapose this framework against the original framework that created these past policies, projects, programmes, housing systems and methodological approaches. To varying extents this framework is implicit and unacknowledged. A scientific approach to evaluation, critique and comparison demands more.

Evaluations, critiques and comparisons presuppose science as emancipatory, as promoting better research and a better society. But the possibility of an emancipatory science depends upon whether we can explicate a normative framework for housing practitioners, a normative framework for housing researchers and a normative framework for those evaluating housing practitioners, housing researchers and themselves as evaluators.

This chapter distinguished three horizons: the horizons of everyday common-sense living, the horizons of theory and the horizon of the subject as subject. The horizons of everyday common-sense living are focused on the immediate issues of living, on things as they relate to us. They vary from culture to culture, from group to group, from housing practitioner to housing practitioner, from individual to individual. Within the horizons of everyday common-sense living our understanding of housing is in terms of its purpose for achieving something else: a standard of living, status, political kudos etc. Thus, as housing is used for different purposes within different horizons of everyday common-sense living, it is understood differently. Conflicts ensue between groups as they each seek to use housing for their purpose rather than another. If we are to resolve these conflicts we need to move to a horizon of theory. Horizons of theory are focused on asking and answering questions which relate things to one another. They vary according to the scholarly tradition within which a housing researcher operates. Within the horizons of theory our understanding of housing distinguishes between what housing is and its various purposes. But scholarly traditions vary and conflicts ensue as to how housing is to be understood. If we are to resolve these conflicts we need to move to a horizon of the subject as subject. This horizon shifts away from the products of the horizons of everyday common-sense living and the products of the horizons of theory, towards a heightened awareness of the subject in action. It focuses on what is prior to these products, with the intending subject, with the operative dynamics within the subject – the dynamic of the desire to understand, the dynamic of the desire to create something worthwhile, the dynamic of sociality etc. These dynamics underpin our learning, our researching, our doing, our living together. Each is constituted by a set of related elements which must occur for the dynamic to reach its term. In this way, an explication of these dynamics provides norms for understanding, for creating something worthwhile and for sociality. It also provides the grounds for meeting the Hegelian requirement of not only accounting for one view, but also accounting for contrary views.

Dialectic is concerned with the decisions we make as housing practitioners and as housing researchers. Dialectic places a demand on housing researchers to be explicit about the horizon within which they operate. It rejects the view that evaluative frameworks are merely arbitrary, simply a matter of our inheritance, of the family, group and culture into which we were born. It rejects the view that it is enough for evaluators to make explicit their beliefs and their values as if that provided an adequate framework for evaluation. From the decision we make as to the horizon within which we will operate, others proceed.

The horizon of the subject as subject opens up a horizon for evaluating the multiple horizons of everyday common-sense living and the multiple horizons of theory. It is the horizon within which Dialectic as a method seeks to resolve fundamental conflicts between housing practitioners operating within the horizons of everyday common-sense living, between housing researchers operating within some mixture of the horizon of everyday common-sense living and the horizons of theory and between housing practitioners and housing researchers. It is an ongoing process of evaluation which recycles the past, which reaches for the best in the past by integrating that best in a larger framework and which searches out the best of the past as a pointer to the future. Dialectic is an encounter in which a housing researcher allows themselves to be challenged by the achievements of the past. It is a challenge because it reveals something about the human spirit and so something about themselves. It challenges the researcher to integrate this new revelation into their lives. It is a way of 'allowing one's living to be challenged at its very roots by their words and by their deeds…[E]ncounter is the one way in which self-understanding and horizon can be put to the test' (Lonergan [1972] 1990: 247).

Dialectic has two movements. In the first movement, the researcher appreciates what has been achieved through the movement forward of history. It appreciates what has been achieved in housing and in housing research, and seeks their further development – something better. In the second movement, the researcher critiques the past state of affairs by grasping its limitations. It is within this second movement that the researcher notes the gap between what was particularised in history and other possibilities that were better particularisations. The researcher can note how what was particularised served the interests of particular persons or groups or, more profoundly, the operative cultural presuppositions. It identifies those practices that are ideologies which aggrandise certain individuals and groups and consequently alienate other individuals and groups. It is a precursor to repairing the damage, the suffering and oppression of alienated individuals and groups, and the alienation of dominant individuals and groups.

Dialectic proposes a shift in evaluation studies, socio-historical critique, comparative studies and methodological critique: from an implicit, presupposed, unchallenged and unacknowledged alternative horizon to the self-luminous horizon of subject as subject; from a generalised critique of persons or groups to an understanding of why this person or group expressed themselves in a particular way in terms of the dynamics of the subject as subject. On the one hand, this understanding affirms what is worthwhile in what they are doing; on the other hand, it reveals the inadequacies of what they are doing due to the interference of one dynamic of the subject with another.

The functional specialty Dialectic proposes a shift in evaluation/critique *from* simply pointing to the limitations of some state of affairs (whether that be an existing policy, program or project, or the exercise of dominating power by individuals and groups acting in their own interest as they aggrandise themselves by alienating others, or the unresolved problems within a housing system in a country, or the inadequacies of housing research), *to* resolving the conflicts of the past by answering in an integral way the evaluative/critical question: 'what is the best of the past?', 'what has been the most worthwhile direction in the history of housing and the history of housing research?' Where evaluation and critique currently operate within the horizon of everyday common-sense living, Dialectic presupposes not just a displacement to the horizon of theory but also a displacement to the horizon of the subject as subject. Dialectic is an encounter with persons that challenges and invites us to integrate new understandings and new ways of living.

Notes

1   See http: //sociology.berkeley.edu/faculty/wacquant/.
2   This more precise understanding of the functional specialty Dialectic is loosely based on (i) Chapter 10 of *Method* (Lonergan [1972] 1990) (ii) Subject and Horizon (Lonergan [1957] 2001b) and (iii) Horizon, History, Philosophy (Lonergan [1957] 2001c). These were supplemented by the writings of McShane (2004a, 2004b, 2007a, 2007b).

3   See Ricoeur (1970: 32–35).
4   On this basis, a researcher can then make a *further* judgement as to whether theorising has occurred and whether it is worthwhile, and a decision as to whether theory will inform their research.
5   As a result, this discussion has few references to academic literature. It points to aspects of research and theory. Support for these pointers lies not in what other people say but rather in attending to and verifying our own internal processes. See the discussion below on Dialectic and housing researchers within the horizon of the subject as subject and the references at Note 15.
6   A theory of theory provides us with a heuristic structure for developing all theories. A social theory such as a theory of sociality will have this general structure, but is relevant within a more defined range, viz. theories concerned with social relations. The core problem of sociality, our intersubjectivity, is explored by Gibson Winter (1966). He begins with the 'I' and 'me' problem as formulated by George Herbert Mead (1938) and draws on the work of Alfred Schutz (1972).
7   Such an understanding of theory would overcome the apparent dichotomy between fact and value that has plagued the history of the social sciences. See James Sauer (1995) for a lengthy discussion of this dichotomy. Sauer notes that Robert Heilbroner and Joan Robinson 'argue that every economic theory implies an understanding of the nature of the economic actor (homo sapiens). The value element enters with assumptions about behaviour and without these assumptions no conclusions can be drawn from social facts' (1995: 44). (Sauer is referring to Heilbroner (1973) and Robinson (1962).)
8   'Horizon' is a term commonly used by hermeneutic and phenomenologist philosophers such as Dilthey, Husserl, Gadamer, Jaspers, Marcel and Sartre (Bunnin and Yu 2007: 311). It has various meanings, and the following paragraphs spell out its particular meaning here.
9   Another way of understanding our horizon is in terms of the questions we can ask. At any moment in time, we can distinguish between three fields, between 'the questions we raise and answer, the questions we raise but cannot answer, and the questions that we neither raise nor answer' (Lonergan [1957] 2001b: 283; see also Lonergan [1957] 2001c: 298–299). Within these three fields, the former is what we have come to understand, what we have taken an interest in and followed through; the latter is what we cannot even ask questions about, what lies completely beyond our awareness, what we are completely ignorant of; in between is what we have yet to understand, what is of some interest, what we know we could investigate, what we know is part of the common fund of knowledge. As researchers, our horizons are the boundary between the questions we raise but cannot answer and the questions we neither raise nor answer, nor have any awareness of. It is the latter, our complete ignorance of vast areas of our lives, that can wreak the most havoc in research (Melchin 1998: 29–30). It is the former that orients us towards future areas of research.
10   See also Lonergan's discussion of the 'general bias' of common sense ([1957] 1992: Chapter 7, Section 8, 250ff).
11   It was in response to this problem of competing understandings of what people assumed they knew that shifted Socrates (who asked the Athenians, 'what is courage?') and Aristotle towards theoria. See also John Dewey's distinction between common sense and science in Chapter 4 of *Logic: The Theory of Inquiry* (1938).
12   See Lonergan (Lonergan [1957] 1992; [1972] 1990: Chapter 1), McShane (1989) and Flanagan (1997).
13   'Ontology, a logos of being, occurs when you start talking about being, but the ontic is the being that is prior to your talking about it' (Lonergan [1957] 2001c: 311).
14   See Hegel's *Lectures on the History of Philosophy* ([1805–1806]1955: 1, Introduction, esp. 1–49) referred to in Morelli (1995: 379 fn2).

15 This discussion is introductory and skates over some very difficult issues. For a more extended and technical discussion, see Lonergan (1967, [1957] 1992, [1972] 1990), Melchin (1999, 2003) or, more popularly, McShane (1974) and Benton et al. (2005).

16 Here I have largely focused upon the dynamic of the desire to understand as this is of principal concern to researchers. But other dynamics also need to be explored for a fuller understanding of this dynamic of the desire to understand – biological, aesthetic, artistic, dramatic, immediately practical, mystical etc. (Morelli 1995; Lonergan [1957] 1992: 204ff, 410–411).

17 On how other dynamics can interfere with the dynamic of the desire to understand, see Lonergan ([1957] 1992) and his discussion of dramatic bias (the interference of the dynamic of the psyche) (p.214ff), of individual bias (the interference of the dynamic of spontaneity) (p.244ff), of group bias (the interference of the dynamic of sociality) (p.247ff) and of common sense (the interference of the dynamic of immediate practicality) (p.250ff).

18 An image that may help to grasp the radical transformation in the shift to the horizon of theory and to the horizon of the subject as subject is the metamorphosis of the caterpillar into the butterfly or the tadpole into the frog. Both the caterpillar and the tadpole 'contain' the capacity for transformation into something very different.

19 Testament to this is their long, slow and difficult emergence, with each emergence marking a major shift in human history (Lonergan [1972] 1990: 85–99, 257–262; McShane 1984; Drage 2003; Shute 2013a, 2013b; see also Jaspers 1953).

20 Even the development and articulation of the three horizons outlined above are products of a long history of reflection on thinking. Chapter 8 points to the transformation of Foundations as philosophy shifts from metaphysics to epistemology and from epistemology to the dynamic of the subject as subject.

21 This complex process of the method of dialectic is discussed summarily by Lonergan ([1972] 1990: 249–250). McShane (2004a) has sought to explicate this page in his Sofdaware series.

22 See also Raymaker (1977) for a critical review of Winter's book.

8 Implementation and the future questions

Dialectic **Foundations**

History **Policies**

Interpretation **Systematics**

Research **Communications**

Let us suppose that a writer proposes to communicate some insight A to a reader. Then by an insight B the writer will grasp the reader's habitual accumulation of insights C; by a further insight D he will grasp the deficiencies in insight E that must be made up before the reader can grasp the insight A; finally, the writer must reach a practical set of insights F that will govern his verbal flow, the shaping of his sentences, their combination into paragraphs, the sequence of paragraphs in chapters and of chapters in books. Clearly, this practical insight F differs notably from the insight A to be communicated. It is determined by the insight A as its principal objective. But it is also determined by the insight B, which settles both what the writer need not explain and, no less, the resources of language on which he can rely to secure effective communication. Further, it is determined by the insight D, which fixes a subsidiary goal that has to be attained if the principal goal is to be reached. Finally, the expression will be a failure in the measure that insights B and D miscalculate the habitual development C and the relevant deficiencies E of the anticipated reader.

Bernard Lonergan in *Insight: A Study of Human Understanding*
([1957] 1992: 579)

'How can I tell what I think till I see what I say?' So exclaimed the old lady to her nieces who accused her of being illogical. This brief anecdote, originally by André Gide but referred to by E.M. Forster in one of his reflective lectures on English literature ([1927] 1974: 71), is one of the difficult issues we face in this chapter. A creative act, one which produces something new, is not simply a logical process. Yes, there is something we grasp that guides the creative process but it is only in the saying and doing that this something takes shape. Indeed, it is only through its expression (whether an inner expression or an outer expression) that we can then ask: is this what I think, is this what I want?

The previous four chapters discussed the research functional specialties – Research, Interpretation, History and Dialectic. Their function is to understand the past. This chapter will discuss the implementation functional specialties: Foundations asks the transformative/visionary question; Policies asks the policy question; Systematics asks the strategic question; and, Communications asks the practical question. Their function is to implement something new in history.

Even before the days when Marx sought to shift philosophers from interpreting history to directing history, men and women had not only been appalled by poverty, by injustice and by the abuse of authority and had sought to right these wrongs, they also aspired to something more for themselves, for their children and for future generations and sought to develop solutions to current problems.

The growth of the social sciences in the twentieth century brought with it not only analysis of social problems and proposed solutions but a strong focus on implementation, with new fields of study opening up such as policy analysis, social administration, urban studies, strategic studies, future studies and implementation studies (Hill and Hupe 2002). New techniques such as forecasting, scenario analysis, foresight (Tegart 2003; Voros 2003; Burke et al. 2004; Burke & Zakharov 2005), transition management (Loorbach & Rotmans 2010), phronesis (Flyvbjerg 2001; Schram & Caterino 2006; Marston 2008) and evidence-based policy (Davies et al. 2000; Pawson 2002; Young et al. 2002; Productivity Commission 2010a, 2010b) have become more commonplace.

I begin this chapter with an initial view of implementation through a phenomenology of research and with a review of five genres of housing research. There follows an extended analysis of some work undertaken by Andrew Jones and Tim Seelig on the linkages between research and policy. This work illustrates both the structure of implementation (and its division into four functional specialties) and some of the weaknesses of current understandings of research-policy or research-implementation linkages.

An initial view of implementation

Housing research is about discovering something new, some new meaning or significance that we had not previously known. It may also lead to some new discovery about ourselves, about doing research, about our sociality, about the world around us and about our society and culture. Attendant upon any discovery, particularly a major discovery, is a question for the housing researcher: what do I do about what I have discovered?

A discovery, particularly a major discovery, invites a decision to act in some way. Do I decide to ignore what I have discovered? Do I decide to reject it because I will have to act in a different way? Do I decide to embrace it and live differently or act differently or write differently?

Some discoveries may fit with our current understanding and demand very little response. On the other hand, housing researchers, in response to other discoveries, might (i) further investigate some housing issue or research issue that confronts them or puzzles them or (ii) communicate their discovery to others or (iii) propose a solution to some housing issue or research issue or (iv) change their approach to housing research. Each situation invites us to do something different from what we had previously done – investigate a different issue or problem, write different things or do different things. In all these different situations, the question for the researcher is: What do I do? What do I do to understand this housing issue or problem? What do I do to communicate the results of my housing research or my discoveries about research? What do I do to bring about a solution to a housing problem or to a research problem?

As researchers in these four situations we are pivoting between what we have discovered and what we are going to do in response to this discovery. Just as there is a structure to researching – asking and answering an empirical question, asking and answering a theoretical question, asking and answering a historical question, asking and answering an evaluative/critical question – so too there is a parallel structure to implementation.

First, I have not only made a discovery but have decided to act on this discovery. So, I begin with a decision to integrate this new discovery into myself. My discovery and my decision changes who I am, changes my living in the world, my interests, concerns, history, skills, understanding, desires etc. The discovery may be one about myself, about myself as a researcher, about myself in relation to others, about the world around me, about my society and culture, about housing. It may result in a small change or in a radical change. This new sense of who I am forms the background or horizon for further decisions.[1]

Second, all four situations described above involve some affirmation: Yes! This discovery is really worthwhile doing something about, it is really worthwhile creating something new here. In creating something worthwhile, we are creating some value. So, it is worthwhile creating a better understanding of some housing issue or problem, either for myself or in someone else, it is worthwhile creating better housing; it is worthwhile creating a better understanding of research, either for myself or in someone else.

Third, they involve a consideration of various strategies – investigating, communicating or bringing about – through which I could do something, create something new of value. These strategies are processes through which something new is created. Investigating, communicating or bringing about involve adapting commonly accepted or taken-for-granted routines in ways which will create or realise a value. These routines form the context within which my discovery will be implemented – we take these routines, whether of language or process, and adapt them so that they better accord with what we have discovered is worthwhile.

Fourth, they involve selecting one of these possible strategies and working out the particular details that actually create or realise this value. It is not until we reach this point that what we have discovered is realised.

So, to come back to the old lady's problem noted at the beginning of this chapter: 'How can I tell what I think till I see what I say?' As Charles Hefling (2000: 148) notes in his reflection on the process of writing a book, implementation is:

a long drawn-out process of knowing what I think by seeing what I say. All the various activities that make up the craft of authorship, everything from outlining chapters to weighing the relative merits of words that are nearly synonymous, are aspects of the slow metamorphosis through which an idea turns into a book.

On an initial view, then, the implementation functional specialties (Foundations, Policies, Systematics and Communications) are concerned with aspects of what-to-do-questions. The context for a what-to-do-question is the discovery of something about the past which invites some expression by putting in place something new or different from our current expressions or practices.

Implementation in housing research

Here I will explore five genres of housing research that are concerned with implementation: research methods, aspirations/visions for housing, policy development, housing strategies and plans, and applications.

Research methods

In Chapter 4 I discussed research methods in relation to how we go about primary research. In the previous chapter on Dialectic, mention was made of a genre of housing research concerned with a methodological critique of housing researchers and the basic positions that inform their research. Besides methodological critique, there is the positive function of housing researchers thematising the grounds for further research, outlining their own stance for doing research and what they understand themselves to be doing as they do their research.

Most housing research is undertaken within the horizon of everyday common-sense living where questions of foundations rarely arise. Where they do arise, they tend to be dismissed as not practical. The problem is reduced to getting government to act and solve an immediate problem. This is reflected in common refrains or recommendations such as 'build more social housing' or 'increase rent assistance' or 'provide grants to first-time owner-purchasers'. These refrains tend to overlook the complexity of doing just that. For those operating within the horizon of everyday common-sense living (and its attendant politics), questions and critique can be ways of undermining the alternative position rather than understanding the grounds on which the problem is formulated or a solution proposed. This taken-for-granted view of the world underpins the positivist approach to housing.

Social construction research (Kemeny 1992; Jacobs & Manzi 2000; Jacobs et al. 2004) and its variants, housing pathways research (Clapham 2002, 2004) and critical discourse research (Marston 2000, 2002), argue that housing researchers adopt a social constructionist viewpoint:

A constructionist epistemology purports that an individual's experience is an active process of interpretation rather than a passive material apprehension of an external physical world. A major claim advanced by those adopting a social constructionist epistemology is that actors do not merely provide descriptions of events, but are themselves constitutive of wider policy discourses and conflicts. Viewing society and social policy as malleable and subject to power struggles, constructionists do not accept social facts as permanently 'accomplished'. This emphasis on contestation is important in offsetting any tendency by actors to objectify social phenomena or reify abstractions into material realities.

(Jacobs & Manzi 2000: 36)

Social constructionism highlights the active role that social agents have in the construction of their world. Social constructionists anticipate that housing will be understood when the interests and motivations of the participants are understood. Their questions are about how the participants are using housing in their own interests. They understand the current construction of housing as the result of conflicts and power in which the powerful tend to reify it in their own interests. If housing is socially constructed, it is not fixed and permanent. Rather, its current form can be contested and it can be reconstructed in a new form. Implicit within this position is a view of the researcher as a recipient of a world already constructed by the powerful. Their role is to reveal these interests or deconstruct this world and reconstruct a new one.

Julie Lawson (2006: 19) makes very clear her stand with a critical realist epistemology and ontology which

challenges the researcher's view that only that which is observable is what exists. It promotes active acknowledgement of the structured, open and dynamic nature of the object or phenomenon for explanation – important causal dimensions that may or may not be directly observed or recorded... Reality comprises events (and non-events) or 'actuality' that may or may not be experienced or recorded, and further, is influenced by emergent possibly unobservable relations with a tendency to produce certain outcomes under certain contingent conditions.

Lawson anticipates that she will be able to explain the divergence of housing solutions between Australia and the Netherlands as causal mechanisms formed from the necessary relations between agents which are 'actualised in the context of other sets of interacting contingent relations' (2006: 28). She anticipates that these mechanisms will be understood through abductive and retroductive methods.

Jago Dodson (2007) presents another basic methodological stance, a 'pragmatic discursive methodology' for 'comprehending the production of housing policy discourse'. Within this methodological stance, 'the concern of philosophical pragmatists to avoid absolute conceptions of truth and to emphasise the empirical constitution of social knowledge forms a highly fruitful basis for the investigation of the discursive production of housing assistance policy' (2007: 55).

Milligan et al. (2007) propose 'a model for implementing a proactive, systematic and achievable program of evaluation' (p.2). Referring to Scriven (1991), they define professional evaluation as 'the systematic determination of the quality or value of something. In the context of public policy assessment, evaluation is a form of research that systematically investigates how well a policy, program or project is meeting its objectives' (p.8). They distinguish evaluation from program monitoring and note that

> in most respects evaluation is similar to other research processes. Thus, to achieve good practice in evaluation research requires the same principles of enquiry, conceptual clarity, methodological rigour, verification techniques and codes of conduct followed for all forms of research…One distinguishing feature of evaluation research is its timing. The research process is usually conducted after the policy or program has been implemented…
>
> (Ibid.)

They, then, go on to develop a realist approach to evaluation with its three elements of Context, Mechanisms and Outcomes (C-M-O), supplemented by additional elements taken from other evaluation traditions.

Other authors propose more specific frameworks for evaluation: Spiller Gibbins Swan (2000) and Ruming (2006) for public housing renewal; Judd and Randolph (2006) highlight the importance of qualitative methods; and Walker et al. (2002, 2003) and Rogers et al. (2005) for indigenous housing programs.

Aspirations or visions for housing

In research and politics, sometimes implicit, are expressions of the aspirations and visions for housing. Some Australian examples include: Oswald Barnett's vision of new housing to replace the slums of inner Melbourne (Barnett & Burt 1942; Barnett et al. 1944); the 1943 Commonwealth Housing Commission's vision of social housing as a broad based tenure (1944); Jim Kemeny's vision of a dominant cost-rent social housing sector (1981, 1983); the Industry Commission's vision of a more efficient public housing system targeted at those most in need (1993a); and the vision of Ecumenical Housing, ACOSS and National Shelter for an expanded, diverse social housing sector (Bisset & Victorian Council of Churches 1991; Ecumenical Housing 1997; National Shelter 2001, 2004, 2009; ACOSS 2008).

These differing visions for the future express the basic stances of organisations and researchers towards housing. Each incorporates not only some understanding of housing within the broader context of its role in society, its role in the development of persons, its role in the economy, but also of its possibilities for providing better living conditions for people and better relations between people. It is within the context of these visions that future policies and directions are formulated and implemented.

Policy development

Policy development is a strong theme within housing research. Governments, housing policy-makers and housing managers have demanded that housing research is policy relevant. It is this direct link between research and policy that characterises much research (Davison & Fincher 1998; Winter & Seelig 2001).

In Australia, for example, policy development covers every aspect of social housing from its initial development through the key periods of its transformation even as interest in policy waxed and waned as problems surfaced in Australian politics. In the 1930s and early 1940s due to the housing shortage in the wake of the Depression, advocates and governments proposed the development of public housing and the formation of SHAs. In the 1970s, reports by the Commission of Inquiry into Poverty (1975), the Priorities Review Staff (1975) and the Australian Institute of Urban Studies (1975) reviewed the basic directions of housing policy. They were highly critical of these directions and proposed new policy directions such as changes in the eligibility for public housing, the development of housing co-operatives, the separation of tenancy management and asset management, and a market rent policy. In the late 1980s and early 1990s governments worked through the implications of economic rationalism for housing. The Australian government initiated a series of reviews in order to work out new policy directions: the National Housing Policy Review (1989; Econsult (Australia) 1989; Flood 1989) proposed funding for social housing through grants rather than loans; the National Housing Strategy (1992)[2] proposed an expansion of social housing; the Industry Commission public inquiry into public housing (Industry Commission 1993a, 1993b) proposed market rents, a separation of tenancy management and asset management, and a new housing subsidy to replace rental rebates. As input into these government reviews, parliamentary inquiries and government budget processes, community organisations such as National Shelter, ACOSS, VCOSS and Ecumenical Housing advocated for increased funding for social housing and for a diverse social housing sector.[3]

As well as policy proposals which sought to improve the broad framework, others sought to improve particular aspects of social housing. Mike Berry and Jon Hall made proposals on financing social housing (Berry et al. 2001; Berry & Hall 2002) and, on reversing the trend towards increasing deficits in public housing (Hall & Berry 2004) and community housing (Hall & Berry 2009). Kath Hulse and Terry Burke made proposals on access to social housing (Hulse et al. 2007) and on housing allocations (Hulse & Burke 2005). Sean McNelis made proposals on the rental system (McNelis 2006), on high-rise housing (McNelis & Reynolds 2001) and on the quality of life for older persons (McNelis et al. 2008). Keith Jacobs and Kathy Arthurson (2003) proposed changes to allocations, to tenancy conditions, to communication between tenants and housing providers, and to working relationships with external agencies in order to intervene more effectively in anti-social behaviour on public housing estates. Vivienne Milligan and others made proposals on the provision of affordable housing by not-for-profit organisations by providing adequate capital funds, by developing a more comprehensive policy and regulation regime, by being more proactive in planning reform and land supply processes, and by building capacity within not-for-profit housing associations through training and evaluation (Milligan et al. 2009).

These policy proposals are but a small sample of those that envision a new direction for the future.

Housing strategies and plans

Policies are a formulation of some future direction. They have yet to be implemented and their implementation requires housing researchers and practitioners to consider the context within which they will be implemented. Housing strategies and plans are the genre of research which consider how policies will be implemented.

In Australia, for example, governments as well as organisations have developed housing strategies and plans. At the national level, after extensive investigations and consultations, the National Housing Strategy (NHS) (1992: 18) outlined a strategy for achieving two broad objectives: affordable housing and appropriate housing. These were set within the broader context of four imperatives for reform: demographic and related social trends; social justice; sustainable economic achievements; and the integration of housing and urban issues with environmental considerations (pp.7–14). In this context the NHS set out three strategic objectives as well as a number of strategies for achieving each of these objectives, and a set of policies and required actions related to each strategy (pp.43–106). This agenda for action

> represents, in the view of the NHS, the future work required for concerted housing and urban reform. It provides an integrated mechanism for the introduction of effective and sustainable reform across a wide front. Pursued with commitment and vigour, which will be essential for success, its implementation will contribute to: housing outcomes…and the future viability of Australia's towns and cities. (pp.29–30)

State government housing departments have developed strategies for the implementation of social housing (Victoria Housing and Community Building 2004; Building South Australia 2005; Housing NSW 2007, 2008). Housing NSW (2007), for instance, outlines a five-year plan for community housing. This plan is located within the context of a vision which 'places people at its centre' and 'offers housing for more people, tailored to their needs, in a way that strengthens local and regional communities' (p.5). Further, it locates the plan within the broader context of: the high and complex needs of tenants; the impacts of demographic change and increased targeting on revenue; creating sustainable communities; increased demand for affordable housing options; and accessing additional funds (p.10). It is in relation to this context that Housing NSW outlines its objectives and related actions for achieving them (pp.13–23).

In a similar vein, some local governments have developed housing strategies (McNelis et al. 2005b, 2005c; City of Melbourne *c*.2001, 2006; City of Boroondara 2007; City of Sydney 2009; City of Knox 2010). The relationship between local affordable housing policies and the context within which they are achieved is highlighted diagrammatically in a Shire of Yarra Ranges Housing Strategy Issues Paper (McNelis et al. 2005c: 4–8).

The primary concern of these strategies and plans is not with housing policy *per se*, i.e. its future direction, but rather with locating this policy within a broader context, in particular, as we have emphasised here, within the broader goals of government. A strategy takes this broader context into account in how this policy will be implemented.

Applications

The primary concern of strategies and plans is the contexts within which a policy is to be implemented. Some housing research, however, also specifies the actions that an organisation (whether government or housing organisation) needs to take to implement a particular strategy. The level of specification varies. For instance, in Australia, the National Housing Strategy (1992) specifies the actions to be taken by governments and others as well as some timelines but generally does not specify the details of who is responsible, what skills they would need, what processes etc. would need to be established; whereas the City of Sydney Draft Affordable Rental Housing Strategy (2009) not only specifies actions to be taken but also who is responsible, what resources are required, the time-frame and the external partners who will be involved.

Housing organisations articulate the commonly accepted ways of doing things within their organisation through operational documents or policies and procedures. Some operate without such documents and to some extent most operate against a background of informal rules, spontaneous agreement or informal learning on how things are done. When differences or conflicts arise and an organisation wants consistency, however, it will document these processes. For instance, the Victorian Office of Housing (the public housing authority) through a series of manuals, has working policies and procedures on calculating household rents (2009), rental arrears (2006–2010), acquisition of properties (2008–2010b), eligibility and allocations (2010). These manuals briefly outline the current policy and then work through how this policy will be applied in particular circumstances.

Housing researchers not only undertake primary research, develop conceptual frameworks, undertake evaluations and comparisons with other countries, propose policies and develop strategies and plans, they also express their results and findings through writing books, articles and papers and through presentations in conferences and workshops etc. For example, as a way of keeping policy-makers informed of current research, AHURI regularly produces Research and Policy Bulletins which outline the key findings of research and their policy implications. For housing researchers, these books, articles etc. are different forms of applications, different practical ways of communicating their discoveries to other housing researchers and practitioners.

The implementation functional specialties in research-policy linkages

As a starting point for developing a more precise understanding of the implementation functional specialties, I will review two related papers by Andrew Jones and Tim Seelig as part of an AHURI project on research-policy linkages in Australian housing: a Discussion Paper (Jones & Seelig 2004) and an Options Paper (Jones & Seelig 2005). These open up a central issue for housing research: the nexus between research, on the one hand, and its implementation in new practices, on the other.

Their broad aims were: 'to develop a systematic framework for understanding the relationships between social science research and public policy, and to use this understanding to enhance the linkages between social science research and policy practice in the Australian housing system' (Jones & Seelig 2004: 1). In the Options Paper, these broad aims are outlined more specifically:

> to propose for discussion a strategy, a set of broad approaches and a series of specific options to enhance linkages between research and policy in Australian housing, with particular reference to the role of the Australian Housing and Urban Research Institute.
>
> (Jones & Seelig 2005: 1)

These papers not only illustrate the implementation of new practices that will enhance research-policy linkages, they also illustrate implementation in two other areas: the implementation of a project, in this instance, a project on research-policy linkages, and the implementation of research-policy linkages over the past thirty years. The following briefly summarises these three sets of implementation.

Implementing the project

Jones and Seelig envisage a project through which they aspire to understand research-policy linkages and propose changes to current practices. This aspiration is the first step in implementing a project which is followed by three further steps: deciding their aims; deciding which strategies they anticipate will achieve these aims; and deciding which particular activities to undertake.

Their *aims* are outlined in the introduction above. Their project methodology (2005: Appendix 1) specifies three *strategies* to achieve these aims:

1   A review of *international literature* on research-policy linkages: 'a comprehensive listing and analysis of the academic and policy literature pertaining to research-policy linkages during the past three decades' (2005: 44)

2 *Workshops*: the first with housing policy researchers and a second with housing policy practitioners, and

3 *Participant observations* by members of the study team.

Prior to actually undertaking *activities*, the authors made decisions as to what in particular they were going to do, such as: where, how, how many and over what time period they would locate particular books, articles, papers etc. on research-policy linkages, and select and review those relevant to the project; where, how, who and how many etc. would participate in the workshops, how they would be organised, what questions would be asked; where, how and who would observe which group of people.

In reflecting upon their work, the authors note that the project operated on another level besides these external processes of formulating aims, formulating strategies, deciding upon particular activities and undertaking these activities. A distinctive aspect of the project is 'its aspiration to utilise this literature as a stimulus for the process of reflection by housing researchers and policy makers on how they can best work together to achieve improved housing outcomes for Australians' (2004: 1). They describe the research process as 'explicitly action-oriented' and 'designed in three stages: analysis, reflection and implementation' (2004: 1). The 'analysis' stage involves a 'critical, conceptual and applied analysis' of the international literature on research-policy linkages. The 'reflective' stage involves Australian housing policy-makers and housing researchers reflecting upon the results of this international review. The implementation stage involves actions which improve the research-policy interface in Australian housing policy.

Implementing research-policy linkages: themes and typology

In the 'analysis' stage, the international literature review provided a framework for structuring a discussion about research-policy relations around four primary questions/themes:

1 *Aspirations and expectations* to more effectively link research and policy: what are our aspirations and expectations concerning relations between research and policy in the context of Australian housing?

2 *Theories and models* of research-policy relations: how should we conceptualise research-policy relations in Australian housing as a basis for the enhancement of these relations?

3 *Structures and processes* to institutionalise research-policy links: how should we structure research-policy relations in Australian housing as a basis for enhancement of research-informed policy?

4 *Practices* to achieve more effective engagement between researchers and policy-makers: what repertoire is best suited to enhancing research-informed policy in Australian housing?

These four inter-linked themes/questions 'represent the central concerns of this literature' (2004: i). They move from an aspiration/expectation (a central principle or idea) outwards through models, structures and processes towards practices. The more central guides and directs the less central, so a decision as to aspirations and expectations guides and directs models which in turn guide and direct structures and processes which in turn guide practices. As the authors note, the debate about aspirations and expectations between 'champions', 'sceptics' and 'reformers' 'sharpens consideration of what research-informed policy really means in practice'.

This move from a central principle or idea towards practices is a theme not only throughout the Discussion Paper,[4] it is also a recurring theme within the literature. For instance, in the discussion of models of linkages between research and policy, the authors develop a typology of research-policy linkages. They identify three models – the engineering model, the engagement model and the enlightenment model – which 'present alternative conceptions both of how research actually links to policy and of how it should link' (2004: 14).[5]

In the *engineering* model, the researcher is a technician in the service of the policy-maker: the policy-maker defines the problem to be solved; the researcher solves these particular policy problems through applied research. In the engineering model, the result is practical solutions to problems.

In the *engagement* model, the researcher and the policy-maker are collaborators, together working out the best policies/directions within a messy and changing world: the researcher brings their 'distinctive knowledge, skills and values of the social sciences'; the policy-maker brings their particular knowledge, skills and values of the particular situation. In the engagement model, the result is policies which propose to address the current political situation.

In the *enlightenment* model, the researcher and the policy-maker live in different worlds and relate to one another indirectly. The researcher operates independently of the policy-maker and is driven by the perspectives of their academic disciplines. The researcher is a detached, sceptical social critic who seeks to clarify issues and inform wider public debates. The policy-maker may or may not take up the challenge posed by the researcher, may or may not rethink their assumptions and frames of reference.

Jones and Seelig have highlighted the differences between these three models of research-policy linkages. From another perspective, however, they could be understood within a larger framework of movement from central principle or idea outwards to practice: the enlightenment model is largely concerned with new principles or ideas for policy; the engagement model is largely concerned with new structures or processes through which policy is achieved; the engineering model is largely concerned with new practices whereby policy is actually implemented.

Implementing research-policy linkages in Australian housing

The results of this project are outlined in the final Options Paper. The authors proposed a strategic direction for AHURI in three parts: first, a strategy of engagement whereby AHURI operates as a 'network organisation'; second, eight general approaches to enhance research-policy relations through a strategy of engagement; third, twenty-four specific options (as specifications of the eight general approaches) which were required to put the strategy of engagement into practice (2005: 1–2).[6]

The endpoint of the project is a set of twenty-four specific options for AHURI. Our question here is: how did the authors come up with these practical options? What did they do? How did they justify their advocacy of these options?

At one level, the explanation lies in the methods used by the authors to reach these results. In the 'analytical' stage, the international literature review provided ideas. So the discussion of practices, structures and processes, theories and models in the Discussion Paper become 'proposals' which were tested, refined, expanded and confirmed through the workshops and became the specific options, approaches and strategic direction in the final Options Paper. The four inter-linked themes/ questions provided a structure for key questions (2004: Appendix 1) for the workshops which focused specifically on Australian housing.

The development in their thinking as a result of 'the reflective stage' is illustrated by the changes in the specific options from the Discussion Paper to the Options Paper: through discussions within the workshops, five 'broad sets of prescriptions for enhancing research-policy relations' (2005: 16) identified in the literature were expanded into 'eight broad 'approaches' to closer and more effective engagement between the research and policy communities'.

In summary, then, the grounds on which the specific options for AHURI were put forward and justified are twofold: (i) the ideas derived from the literature review and (ii) the refinement, expansion and confirmation of these ideas as they were tested through the workshops and participant observations.

What is less obvious, however, is the role(s) each of these core activities played in the formulation of the overall strategy, general approaches and specific options. In exploring this role, it is easier to begin with the endpoint, the specific options.

Specific options: towards new practices

The aim of the specific options is to *improve* the research-policy interface in Australian housing. So, something better than what was there previously. Whether these options will or will not improve the research-policy interface has been tested through the methods outlined above: the review of international literature; the workshops and the participant observations. While these options may be the subject of some debate, their specific content is not the issue I wish to consider here.

Rather, I would note five things in relation to these specific options:

1   The options are geared around the particularities of AHURI, in this time and place with certain opportunities and possibilities for change. Thus, the discussion of the specific options in the Options Paper (pp.17–37) is accompanied by an analysis of the situation in which the change will occur.

This includes some history in relation to the particular issue they are intending to deal with. They are geared towards a 're-formation' or 'reintegration' of current and possible activities in a new way.

2   As noted by the authors, the implementation of these options will depend upon 'the commitment and support from research centres, housing authorities, and individual researchers and policy practitioners, as well as from AHURI' (2005: 39). In other words, these options are still conditional. Moreover, while very specific in nature, they have yet to be implemented in a way which will fit with and be acceptable to external agencies.

3   The options are a series of new actions or practices. The motivations of the various actors, whether they be individuals or organisations, are not relevant.

4   These options are numerous and cover a range of areas. But they are not discrete and unrelated actions or practices. In some way, they are related to one another. The question, then, is: what is it that relates them to one another?

5   Finally, we can note that the authors group these options under eight different approaches.

Approaches: structures and processes

Eight approaches provide the context for the twenty-four options:

1   Adopt and promote engagement as a core model, principle and practice;
2   Engage around the research agenda and research funding;
3   Engage around the conduct of research;
4   Engage around research dissemination and utilisation;
5   Engage in wider policy processes;
6   Promote local level collaboration;
7   Focus on research synthesis;
8   Promote skills development in research-policy linkage.

Each specific option is related to one of these approaches, and the specific options are related to one another insofar as they are part of each approach. Three approaches each incorporate two options, two approaches incorporate three options and three approaches incorporate four options. So, for instance, Approach 2 (Engage around the research agenda and research funding) incorporates Option 5 (Establish a more engaging and participative approach to ongoing development of the AHURI research agenda), Option 6 (Develop a format for the research agenda that is more explicit concerning the diversity of research that will be supported) and Option 7 (Develop an explicit engagement process for the development of new Collaborative Research Ventures) (2005: 22–23, 41).

Each option makes a contribution to the achievement of their respective approach, and together the options in each approach will achieve that approach.

But again, as with the specific options, we can note that these approaches are numerous and cover a range of areas, that they are not discrete and unrelated. Our question, then, is: what is it that relates these approaches to one another?

Models of research-policy linkages – the engagement strategy

As the authors note, 'each approach represents one aspect of an overall engagement strategy, relating to particular organisational structures, stages of the research process, or problematic aspects of research-policy linkages' (2005: 16). The central point here is that the approaches are related to one another as aspects of an engagement strategy for research-policy linkages. It is this strategy that brings them together. It guides the formulation and specification of the eight approaches and, in turn, the formulation and specification of the twenty-four specific options.

For the authors, '[o]ne of the principal findings of the research has been that engagement is fundamental to achieving closer research-policy relations' (2005: 38). Engagement is the principle which determines different aspects or approaches. In turn, different aspects or approaches determined the specific options.

How then, did the authors conclude that 'a strategy of 'engagement'' was 'the most promising approach to achieving effective research-policy linkages' (2005: 4)?

Their review of the literature indicated various theories or models of research-policy relations and how research was utilised in the policy process. After reviewing these, the authors reduced the number of theories/models to three: the enlightenment model, the engagement model and the engineering model. These three models envisage different research-policy linkages. The question confronting the authors is: through which model are research-policy linkages better incorporated?

Such a choice is not simply based upon what has previously happened, i.e. an analysis and understanding of each of the three models, and how policy utilises research in each model. Indeed, what is interesting is that the authors want to hold onto all three models. While they appear to opt for the engagement strategy, they understood it in a way which encompassed the other two models. The authors understood the 'engagement' strategy as encompassing 'both the highly specific research questions typical of the engineering model and the wider, more speculative research questions of the enlightenment model' (2005: 15). In this way, the three models were conflated into one as the engagement model was understood as encompassing the other two in some way.

Aspirations and expectations for research-policy linkages

Prior to the discussion of the three models of research-policy linkages, the authors discuss the ongoing debate over aspirations and expectations for research-policy linkages. They note how these vary and, while recognising that this is an over-simplification, they highlight the differing positions of three broad groups: champions who contend 'that policy should be based on the findings of social science research' and that this 'is self-evident, requiring little if any justification'; sceptics who contend that 'social science-based public policy is a mirage, based on unrealistic assumptions about social science, the policy process and the capacity of researchers and policy makers to work effectively together'; and reformers who neither dismiss nor accept unthinkingly the role of science in public policy but 'suggest the need for new structures and practices if research-informed and research-enhanced policy is to be successfully progressed' (2004: 8).

Most researchers and policy practitioners associated with AHURI adopt a reformist stance. It is the aspirations and expectations of this reformist stance that sets the framework for an engagement strategy, the approaches and the specific options. It is a stance which both appreciates what has been achieved yet critically recognises the limitations of these achievements. It is a stance whose intent is progress in research-policy linkages.

Learning from the past

By focusing on aspirations and expectations, Jones and Seelig find a starting point for implementing new research-policy linkages through AHURI. It is a starting point which looks forward to what can be or what we want to be. To get to this starting point they worked through the history of debates between champions, sceptics and reformers and made an evaluation of them.

Just as champions, sceptics and reformers ask: 'should we seek to realise better linkages between research and policy?', 'is it desirable?', 'is it possible?', so too Jones and Seelig ask themselves the same questions, for 'the goal of research-informed policy...raises fundamental and recurring debates about the nature and purposes of both social science and public policy in democratic societies' (2004: 4).

In seeking to work through these debates on research-policy linkages, Jones and Seelig (2004: 4) propose three key sets of questions:

1   What is the history of the idea that social science research and public policy should be closely linked, and that policy should be research informed? Where does this idea come from and why is it so prominent on the policy agenda at the present time?
2   What are the broad parameters of the debate? What are the arguments and views of those who champion the idea of research-based policy, and of those who raise questions about this aspiration?
3   Specifically, what exactly are we aspiring to when we espouse more effective links between research and policy in Australian housing?

These three questions deal with different aspects of the research-policy linkages. The first deals with the historical experience or what has already happened in relation to these linkages. The second deals with the differing evaluations of this historical experience in view of future directions. The third raises the question as to what exactly are these linkages that we aspire to bring about.

In answering the *first* question, Jones and Seelig point to a whole range of instances where public policies have been based upon social science research and how confidence in such research as the basis for public policies has waxed and waned with perceptions of its effectiveness. The search for better ways to relate research and policy resulted in further reflection and literature on the issue. In particular, the paper reviewed the evidence-based policy movement as it developed in the United Kingdom and the criticisms of their practices. The authors note the pragmatic impulse in this movement, its claims to overcoming ideologically based positions and its compatibility with managerialism. They conclude by highlighting the long history in which research and policy sought to develop closer links, the increasing complexity of those links, the periods of close collaboration interspersed by periods of conflict and the possibilities inherent in the evidence-based policy movement (2004: 4–8).

In answering the *second* question – the debate between the champions and the sceptics about the possibility of research-informed policy – Jones and Seelig outline the arguments for and against research-informed policy. The champions' aspiration for research-informed policy is particularly based on 'a belief in the capacity of social science research to provide answers' (p.9). The sceptics, however, argue that the champions (i) misunderstand the nature of the policy process within politics by underestimating the role of interests, ideology and irrationality in policy and by misunderstanding the nature of the knowledge needed for effective policy making, (ii) over-estimate the capacity of research, (iii) under-estimate the difficulties of developing partnerships between researchers and policy-makers, (iv) ignore the evidence of the limited influence of research and (v) endanger the role of the social sciences in a democratic society.

While the champions point to the positive aspects of research-informed policy, the sceptics point to its limitations and its negative aspects. The reformers' position on research-informed policy seeks 'a more nuanced understanding of the role of research in policy and of research-policy relations' (p.11). 'These 'reformers' accept the basic propositions of the case for research-informed policy, while acknowledging the complexities and constraints of policy and research processes. The stance is one of cautious optimism' (p.12). The reformers take the basic stance of the champions' position and adapt it by taking account of the arguments put forward by the sceptics' position. This reformers' position is much more circumspect with regard to the possible achievements of research-informed policy, while not abandoning the possibility or aspiration altogether (pp.8–12).

In answering the *third* question – what should our aspirations and expectations be? – Jones and Seelig shift to a particular context, viz. the current research-policy linkages within Australian housing, and put a series of questions to researchers and policy-makers (pp.12–14). Their answers to these questions provide the context for their aspirations for future research-policy linkages in Australian housing.

Research-policy linkages and the implementation functional specialties

These two papers by Jones and Seelig illustrate the movement from current research-policy linkages within AHURI to a proposal for better research-policy linkages within AHURI. They begin by learning from the past – considering the history of debates about research-policy linkages and an evaluation of these debates. On this basis, they take a stance which aspires to better linkages; they decide on the best model for expressing this aspiration – the engagement strategy; they decide on a set of approaches for this engagement model; they propose a set of specific options that will put in place these approaches.

The two papers illustrate implementation in three areas: housing researchers implementing a project; the implementation of research-policy linkages in the past; and implementing research-policy linkages in Australian housing.

In what is a thorough review of research-policy linkages, I want to discuss two particular aspects further: first, the role of theory in coming to some aspiration for research-policy linkages and in their implementation; second, the structure of implementation, whether in (i) implementing the project or (ii) the themes/typology that informs past research-policy linkages or (iii) implementing research-policy linkages in Australian housing.

A scientific approach to implementation – whether as a research project, as an understanding of implementation in the past or as implementing some new practices – demands not just research-policy linkages but theory-practice linkages. It also demands a sound basis for structuring that implementation. Our current understanding of the implementation or policy process is dominated by the horizon of everyday common-sense living. We need to make two shifts: first, into a horizon of theory that will give us control of what we are implementing; second, into a horizon of the subject as subject, distinguishing the different questions that emerge in the process.

Shifting implementation into a horizon of theory

Jones and Seelig highlight the importance and difficulty in linking research and policy. But like most policy development within housing studies, the linking occurs within the horizon of everyday common-sense living where both research and policy are understood descriptively.[7] This has the advantage of conveying directly to practitioners something of the importance of research and policy and how they link together. But as discussed in previous chapters, understanding within the horizon of everyday common-sense living makes assumptions about research-policy linkages that incorporate the interests of a person, a group or a culture by not distinguishing between what is being investigated and the role or purposes that it plays. Within this horizon, a researcher is unable to control what is being investigated and they view it from a particular perspective, most often a dominant perspective where it serves their particular purpose.

For Jones and Seelig, this horizon of everyday common-sense living has three implications. The *first* concerns their understanding of research-policy linkages.

At no stage do they address the prior questions 'what is research?' and 'what is policy?' The papers tend to assume that everyone knows what research and policy are. Their problem is the relationship or linkages between them. But an explanatory understanding of research and of policy is essential to grasping the linkages between them. By operating within the horizon of everyday common-sense living, Jones and Seelig do not have an adequate heuristic for considering the varying historical understandings of research-policy linkages by the champions, the sceptics and the reformers and are unable to distinguish between those aspects of these understandings that concern research-policy linkages and those aspects that arise out of their own interests and biases. While they take a stance on research-policy linkages, they are unable to do so on a firm basis. Rather, they appeal to the aspirations of housing researchers. While they distinguish between different models of research-policy linkages – enlightenment, engagement and engineering – and opt for the engagement model, they want to include elements of both the enlightenment model and the engineering model.

The *second* implication relates to the implementation of better research-policy linkages within AHURI. Above, we noted how each of the twenty-four specific options for better linkages related to one of eight approaches which in turn related to an engagement strategy. The assumption underpinning this structuring of specific options, approaches and strategy is that, through the implementation of certain specific options, a certain approach would be achieved. So, for instance, by implementing specific option 15 (Experiment with research-policy workshops on topical issues targeted to specific groups of policy participants) and specific option 16 (Develop a pro-active media and research promotion strategy), approach 5 (Engage in wider policy processes) would be achieved. Further, through the combination of the eight approaches, an engagement strategy would be achieved (Jones & Seelig 2005: 41f). However, it is only upon an understanding of the systematic relationships between certain specific options and a certain approach, and between all the approaches and the engagement strategy, that Jones and Seelig could make these proposals. In other words, they assume that the eight approaches would constitute an engagement strategy and that a group of specific options (whether two, three or four options) would constitute an approach.

As previously discussed in Chapter 5, it is towards an understanding of these systematic relationships that theory is directed. Theory is an answer to a what-is-it-question. A theory of housing, a theory of research, a theory of policy is a complete ordered set of sets of related variable elements that constitute housing, research or policy. While Jones and Seelig may be correct in the way they relate specific options, approaches and the engagement strategy, they may not have included all the significant elements, they may have included some which are irrelevant or insignificant, and they may have included elements which relate to different purposes. Within a horizon of theory, a researcher is seeking to grasp those significant elements and their relationships which constitute research-policy linkages. It is upon this understanding and by grasping that variations in these significant elements and relationships are possible that better research-policy linkages can be achieved.

The *third* implication concerns their focus on persons, groups or institutions. Jones and Seelig speak about researchers, policy-makers, practitioners, government and AHURI, rather than about their activities which form a set of related variable elements that constitute research and policy. For Jones and Seelig, as for most policy-makers, the policy process takes as its starting point a particular group with their particular interests operating within a particular context. Policies, strategies and activities are developed in terms of these particularities.

As we have seen, with the shift to the horizon of theory, a housing researcher is concerned with the set of related variable elements that constitute housing. As we move from understanding the past – in which we differentiate between Research, Interpretation, History and Dialectic – to concern about the future, that same something, housing, is under consideration. Our understanding of the past has brought us to the point where we have gathered the relevant data on housing; we have grasped the set of related variable elements that constitute housing; we have grasped the dynamic that brought about a sequence of actual operating housing systems; and we have evaluated (appreciated and critiqued) this actual sequence (and discerned the extent to which the dynamic of change has served the interests of particular groups).

Our discovery of new meanings and new understandings of the past pose a challenge to us now: what do we do? It is a challenge to move forward, as best we can, on a new basis and in a new way to create a new form of housing. Just as in research our challenge is to grasp the actual set of related elements that constitute housing (rather than understanding housing from the perspective of a particular interest group), so too we have a similar challenge in implementation: to envision and bring about something better for housing *as a whole* rather than something which serves the interests of a particular group. So, rather than a policy process operating within the horizon of everyday common-sense living, what is called for is a policy process operating within the horizon of theory.

Shifting implementation into a horizon of the subject as subject

My discussion of the four types of implementation within the two papers by Jones and Seelig highlights a sequence of decisions, i.e. a structure, within implementation. So, the discussion of implementation of the research project pointed to an aspiration, aims, strategies and particular decisions; the discussion of implementing research-policy linkages in the past pointed to aspirations/ expectations, theories/models, structures/processes and practices, and then to three models operating at different levels: engineering, engagement and enlightenment; the discussion of implementing better research-policy linkages pointed to the strategic direction of engagement, eight approaches that enhance this strategy and twenty-four specific options through which these approaches and this strategy would be achieved.

Now different authors either explicitly or implicitly structure implementation (from determining a direction to new practices) as a sequence of decisions. For example, the final report of the National Housing Strategy (1992) speaks of national goals, objectives, strategies, policies and actions. Obviously, there are differences of terminology. Some authors propose three levels whereas others propose more. The determination of the terminology and the number of levels depends upon the situation, the audience and the complexity of implementation. So, the number of levels and the terms used for each level will vary according to practical demands of conveying some strategy simply and clearly.

Implementation within a scientific approach to housing, however, demands that we distinguish between the levels of the policy process. In the previous functional specialties, this scientific mode has been structured around key questions. I suggest that there is a certain parallel between the first four functional specialties and the four implementation functional specialties. Where the first four functional specialties are concerned with understanding what has already happened in the past, the four implementation functional specialties look forward to implementing something new. In doing so, there is a shift in orientation of the subject from understanding to doing, from understanding the past to looking to the future. We shift from puzzlement to aspiration, from being a researcher confronted with their lack of knowledge to being a researcher implementing a project, from being researchers to policy-makers.

Indeed, the parallel between the research functional specialties and the implementation functional specialties derives not from applying the structure of research functional specialties to implementation. Rather, strangely, it is the structure of implementation that 'imposes' a parallel structure on the research functional specialties. It is because implementation has a certain structure that research has a certain structure. Research is a reflection upon the past, upon implementation that has already occurred.

Just as we distinguished between four different questions that grounded the research functional specialties, so too we can distinguish four similar questions (but in a forward looking mode) that ground the implementation functional specialties. Where a research functional specialty asks an evaluative question, an implementation functional specialty asks the transformative/visionary question; where a research functional specialty asks the historical question, an implementation functional specialty asks the policy question; where a research functional specialty asks the theoretical question, an implementation functional specialty asks the strategic question; where a research functional specialty asks the empirical question, an implementation functional specialty asks the practical question.

Implementation, the second phase of a scientific approach to housing, unfolds in a series of decisions which we have called Foundations, Policies, Systematics and Communications:

> [The movement of the] second phase is from the unity of a grounding horizon, the foundations, through [policies] and systematic clarifications to communications with almost endlessly varied sensibilities, mentalities, interests, and tastes of mankind. Again, in the second phase the process is not deductive, from premises to conclusions. It is a movement through successive and more fully developed contexts: from the foundations, which is just the

horizon, to the [policies], which give a shape within that horizon, to systematics, which clarify, make the shape meaningful, and finally to the communications, which transpose the message to all the different wavelengths.

(Lonergan [1968] 2010: 462)

The following seeks an understanding of each of these contexts as we shift from Foundations in the subject as subject to its multiple expressions in Communications.

Foundations and the transformative/visionary question

Norman Blaikie begins his book, *Approaches to Social Enquiry: Advancing Knowledge*, by noting the theoretical and methodological controversies that plague the social sciences:

> For the past fifty years, the social sciences have been plagued by theoretical and methodological controversies. The demise of structural functionalism in American social theory, and the emergence of new philosophies of science and social science, have been followed by a period of dispute between a variety of theoretical perspectives and approaches to social enquiry. The issues and dilemmas that have been exposed by these contemporary controversies are not new; their ingredients can be traced back to philosophical traditions that were established in the nineteenth century, traditions that have their roots in antiquity. What is new is that social scientists can no longer ignore these issues as being the pastime of philosophers. Social enquiry has lost its innocence. Social researchers must now face up to, and deal with, a range of choices and dilemmas that lead to the use of fundamentally different research strategies and have the possibility of producing different research outcomes.
>
> (Blaikie 2007: 1)

Indeed, housing research as any social inquiry is no longer innocent and must deal with fundamental issues of how researchers approach their work. This difficult territory is the concern of the functional specialty Foundations. It is not good enough, however, to exclaim that

> it is not possible to establish by empirical enquiry which of the ontological and epistemological claims is the most appropriate. The proponents adopt a position partly as an act of faith in a particular view of the world. All that can be done is to debate their respective strengths and weaknesses.
>
> (Blaikie 2007: 25)

In this view, a form of fideism, social science locates its foundations in some arbitrary axiom, self-evident truth, first principle or value stance from which it derives its theories and viewpoints. The future of social science as a science demands that we confront this view head-on. Foundations are not arbitrary.

As we have seen previously, the role of the functional specialty Dialectic is to confront this issue. Here, I am pointing to a problem with the current view on Foundations. Blaikie, in his introduction, proposes three types of research questions: 'what' questions, 'why' questions and 'how' questions (2007: 6). But this is not the basis upon which he proceeds to examine the research process – he does not explore the origins, differences between, relationships between and significance of such questions as they arise. Rather, he ignores these opening questions and goes on to describe a range of different approaches to social enquiry – positivism, critical rationalism, classical hermeneutics, interpretivism and constructionism, critical theory, ethnomethodology, social realism (or critical realism), contemporary hermeneutics and structuration theory. As a result, he never actually 'pins them down' nor grasps the significance of each (what each approach is), why they differ from one another and how they are related to one another. For to do that requires an adequate heuristic, a theory which not only proposes

> the norms or invariants or evidence that recur in every subject independently of his [or her] horizon. It must also account for the fact that de facto people do have different horizons, despite the existence of these norms that recur in everyone. It must account, at least in principle, for the existence of the multiplicity of horizons, and it must have some capacity for explaining the varieties of philosophic opinion that have existed for thousands of years... While it must account for the horizons that occur, it must do so in such a way that it also discredits them. It accounts for them as violations of the norms, humanly intelligible violations of the norms, so that it is plausible – not only possible put plausible – that many philosophers should think in these erroneous ways.
>
> (Lonergan [1957] 2001c: 313)

An inquiry into this range of approaches to social enquiry is the data to be investigated (Research) which challenge the current theory of method and which may give rise to a new theory (Interpretation). This theory provides a heuristic for understanding the vector of the actual sequences of method (History) which is subject to a further question as to whether or not this vector has brought about a development of method or its deterioration (Dialectic). It is in Dialectic that we can confront the disparity between what proponents of each of these approaches say about social enquiry with their actual performance of social enquiry.

In Foundations, however, we shift beyond a consideration of past decisions about method to articulating the basis on which we will operate now.

In the traditional view (as exemplified by Blaikie), foundations are found in first principles, in self-evident truths or beliefs, or in the arbitrary adoption of values. They provide a starting point from which other principles, other truths or beliefs and other values can be derived. However:

> Foundations may be conceived in two quite different manners. The simple manner is to conceive foundations as a set of premises, of logically first

propositions. The complex manner is to conceive foundations as what is first in any ordered set. If the ordered set consists in propositions, then the first will be the logically first propositions. If the ordered set consists in an ongoing, developing reality, then the first is the immanent and operative set of norms that guides each forward step in the process.

Now if one desires foundations to be conceived in the simple manner, then the only sufficient foundations will be some variation or other of the following style: One must believe and accept whatever the bible or the true church or both believe and accept. But X is the bible or the true church or both. Therefore, one must believe and accept whatever X believes and accepts. Moreover, X believes and accepts *a, b, c, d,...* Therefore, one must believe and accept *a, b, c, d,...*

On the contrary, if one desires foundations for an ongoing, developing process, one has to move out of the static, deductivist style – which admits no conclusions that are not implicit in premises – and into the methodical style – which aims at decreasing darkness and increasing light and keeps adding discovery to discovery. Then, what is paramount is control of the process.

(Lonergan [1972] 1990: 269–270)

In the discussion of horizons in the previous chapter, I outlined three different horizons: the horizons of everyday common-sense living, the horizons of theory and the horizon of the subject as subject. I proposed that the horizon of the subject as subject provided a framework for evaluation and critique of the decisions and actions of individuals and groups (and the distorted practices of dominating power and the self-aggrandisement of individuals and groups); of the decisions and actions of housing researchers; and of the decisions and actions of researchers undertaking evaluation, comparison and critique, whether of individuals and groups, housing researchers, or housing researchers undertaking evaluation, comparison and critique. Through a heightened awareness of the subject in action, I proposed that we could thematise or articulate the dynamics of the subject as subject: the dynamic of the desire to understand, the dynamic of the desire to create something worthwhile, the dynamic of sociality and the dynamic of being-in-love.

Dialectic operates on the basis that we learn what is best from retracing our steps. We learn through the long hard-fought ongoing communal process of reflecting on the past and working out as best we can a basis for development and growth (technological development, economic development, political development, cultural development and personal development), a basis for living into the future. By retracing our steps we can make discoveries which *invite* a fundamental transformation of ourselves and our horizon.

Discovery and invitation is one thing. Deciding to live in a new way is another. It marks a shift in mode: *from* learning from the past *to* creating a new future that encompasses both ourselves as subject and our expression in words and actions. It is a decision which pivots between who I am now and who I want to be. It can be a decision not to change, a decision that closes off or restricts my future. Or, it can be a decision based on a discovery of what is best. The foundations of housing research, then, lie in the conscious decision of the subject. It is a decision to implement something new. It is a decision which raises the transformative/ visionary questions: 'will I operate in accord with the dynamics of the subject?' 'will I operate within the horizon of the subject as subject?'

> At its real root…foundations occurs…on the level of deliberation, evaluation, decision. It is a decision about whom and what you are for and, again, whom and what you are against. It is a decision illuminated by the manifold possibilities exhibited in dialectic. It is a fully conscious decision about one's horizon, one's outlook, one's world-view. It deliberately selects the frameworks, in which [policies] have their meaning, in which systematics reconciles, in which communications are effective.
>
> (Ibid: 268)

Foundational reality is a decision about a fundamental orientation of ourselves as subject towards ongoing development, one which embraces future development and how this occurs. Foundations is a decision about our basic orientation or stance on living that will take us into the future. It is concerned with horizons as the framework within which we will make further decisions in the future. This future is not simply a repeat of the past but a leap into something new, a leap which envisages a decision that not only results in a single development in ourselves but rather an ongoing development. So, a

> deliberate decision about one's horizon is high achievement. For the most part people merely drift into some contemporary horizon. They do not advert to the multiplicity of horizons. They do not exercise their vertical liberty by migrating from the one they have inherited to another they have discovered to be better.
>
> (Ibid: 269)

This transformative shift in horizon is not easy because it demands a reorganisation of ourselves as subject. It demands a reorientation in our interests and our living. It demands a change in what we do and how we do it. As housing researchers it demands a change in the data that we attend to, in the questions we ask, in differentiating the activities required to answer one sort of question from those required to answer a different sort of question etc. It demands a change in the groups we work with. Transformative shifts in horizons take time as we learn new skills.

The role of the functional specialty Foundations, then, is to articulate or to thematise how this development occurs. It raises the traditional, and seemingly remote, problems of philosophy, epistemology, metaphysics and ethics as a transformative/visionary question for a housing researcher: 'who am I?' or, to put that same question in another way, 'who do I want to be?' or, 'who am I as a researcher?' Foundations transposes these traditional problems into a personal question for the researcher about the relationship between themselves and their research, about their operations as they go about their research, about themselves as an operating subject.

Up until the seventeenth century, the starting point for most philosophy was metaphysics – the study of the nature of the world, of being and reality. This philosophy took its cues from the great Greek philosophers Socrates, Plato and Aristotle, among others. The questions raised by Descartes, Hume and Kant, among others, shifted the starting point from metaphysics to epistemology, to questions about knowledge and its status, to questions about objectivity. For these philosophers, epistemology grounds metaphysics and, thus, different epistemologies have grounded empiricist, positivist, realist and idealist metaphysics. In the early twentieth century (in the wake of Kant's Copernican revolution, in the wake of Hegel, Marx, Nietzsche and Freud and unresolved questions about objectivity), phenomenologists, hermeneutists and existentialists such as Bergson, Kierkegaard, Husserl, Merleau-Ponty, Buber and Heidegger began exploring subjectivity. The writings of Bernard Lonergan can be understood as a thematisation of this work ([1957] 2001a: 274ff). This thematisation shifted the focus of philosophy from epistemology to interiority or the subject as subject where philosophy takes as its starting point, self-appropriation or a grasp of what I am doing as I know, a grasp of the fundamental dynamics of human knowing and doing. Now, many of the activities that Lonergan refers to have been subject to widespread debate among philosophers of science such as Popper, Kuhn, Feyerabend and Lakatos and have been described in detail by phenomenologists and existentialists. But what is unique about Lonergan's account is that he issues 'an invitation to a personal, decisive act... know oneself'. The shift to interiority or to the subject as subject is a shift to grasping the sets of practices or activities that constitute knowing and, in doing so, distinguishing them from other activities that may interfere with these sets of activities. The key issue, then, becomes one of competent performance. It is the competent performance of knowing that grounds epistemology which in turn grounds metaphysics (Lonergan 1967, [1957] 1992, [1972] 1990).

With the shift to the subject as subject, a metaphysics, an ontology or an epistemology are no longer adequate starting points. Foundations thematises or objectifies the horizon of the subject as subject. The subject is a whole with many dynamic parts. The parts are related to each other and to the whole. It is a dynamic structure, self-assembling and self-constituting, where

> [e]ach part is what it is in virtue of its functional relations to other parts, there is no part that is not determined by the exigences of other parts; and the whole possesses a certain inevitability in its unity, so that the removal of any part would destroy the whole, and the addition of any further part would be ludicrous.
>
> (Lonergan 1967: 222)

Foundations articulates the dynamics of the desire to understand, the dynamics of the desire to create something worthwhile and the dynamics of sociality as the three grounds for ongoing development of housing research. An articulation of the dynamics of the desire to understand distinguishes a complete ordered set of eight inter-related questions. An articulation of the dynamics of the desire to create something worthwhile distinguishes the grounds upon which decisions are made about which research is the best to undertake (not that which is merely pleasing to the researcher). An articulation of the dynamic of sociality reveals the grounds for collaboration among housing researchers. Together these dynamics constitute the grounds for the functional specialties and for Functional Collaboration as a new way of structuring science as a collaborative enterprise.

The functional specialty Foundations, then, seeks an answer to the transformative/visionary question: 'who am I?' or, 'who do I want to be?' or, more specifically, 'who am I as a housing researcher?' The foundations for housing research will not be found in arbitrary axioms, self-evident truths or first principles, they will be found in the dynamics of the subject as subject. This thematisation of the subject as subject articulates the best framework for undertaking housing research into the future, the best framework for the ongoing development of housing research. It becomes a foundation for research when a housing researcher decides to operate within this framework.

From the transformed viewpoint of Foundations we can envision future directions, envision their integration within different contexts and envision the practices through which these future directions can be implemented.

Policies and the policy question

Foundations articulates the horizon within which implementation occurs. The functional specialty Policies is concerned with giving shape to that horizon. It is what is intended in a course of action: 'policy is essentially a *stance* which, once articulated, contributes to the context within which a succession of future decisions will be made' (Friend et al. 1974: 40).

The meaning of any policy statement – a statement of directions – can only be understood in the context of Foundations. So, the meaning of a policy will vary according to whether it proceeds from a horizon of everyday common-sense living, a horizon of theory or the horizon of subject as subject.

But like many policy analysts, Friend et al. operate within the horizon of everyday common-sense living and understand policy as the stance of a government or organisation, a stance within which they will make a succession of future decisions. A stance on housing may lead to decisions about aspects of housing such as taxation arrangements, housing assistance, interest rates, industry support, land supply and land use. Policies of government and organisations reflect their interests and conflict ensues insofar as other organisations oppose these interests.

Here, then, is a fundamental problem with policy as it is discussed in housing research and wider policy research (for example, Hogwood & Gunn 1984; Parsons 1995; Hill 1997, 2003; Hill & Hupe 2002; Fischer et al. 2007). This research discusses policy and policy processes insofar as they are developed by government or organisation. The scope of the discussion of policy is restricted to the interests of government or organisation.

If most policy operates within a horizon of everyday common-sense living – after all, policy is oriented towards practical change – what would policy look like from within the horizon of theory?

A theory of housing is a complete ordered set of sets of related variable elements which brings together not only housing but also the complete range of its possible purposes. It explains an actual operating housing system in terms of a series of approximations where each purpose contributes to the actual system. The discovery of new relevant data, a shift in our understanding of housing, a shift in our understanding of the vectors that motivate the shift from one actual operating housing system to another, a shift in our understanding of what constitutes progress, and our aspiration to make a better housing system raise a question about the future direction and the vector that will motivate the shift from the current system to a new one. An ideology will exclusively focus on one or other purpose of housing that serves the interests of a particular group. What it continually overlooks is the role that housing plays in achieving a whole range of purposes. The challenge of policy, then, is not the adoption of one particular vector, but rather a direction which will promote the *totality* of housing and its purposes.

In History we seek to grasp the vector or vectors that motivated or brought about a sequence of actual operating housing systems. This vector or vectors motivated a change in these systems. The function of Policies is to select, from among a range of possible vectors, the one that will motivate an ongoing development in the housing system as a whole. Policies are concerned with directions into the future. Before we change our practices, before we strategise on how we can change these practices, we envisage some direction. Policies deliberate on possible directions and, from among a range of possibilities, selects a direction for the future. Policies are what motivate this future development in practice. Policies are what we intend. They are a direction or guideline for the future. It is the affirmation of a vector which encompasses the totality of purposes that sets in train a series of further decisions that will emerge from Systematics and Communications.

The functional specialty Policies seeks to answer the policy question: what new operative dynamic or vector will best promote the future development of the current actual operating housing system into future systems? Policies focus on the housing system as whole, on the totality of its purposes. Policies shifts the current discussion of policy development away from a focus on what governments or organisations want or what others want from them – policy in the interests of a particular organisation or group. The focus of Policies is on new directions or vectors that bring about future development. It is a vector through which not only housing will be better realised but also the range of purposes in which housing plays a role.

Systematics and the strategic question

Henry Mintzberg, an 'internationally renowned academic and author on business and management'[8] focusing particularly on strategy within organisations, outlined eight ideal types of strategies along a continuum from strong central management to external imposition as follows:

1 Planned strategies: originate in formal plans of central leadership;
2 Entrepreneurial strategies: originate in a central vision of a single leader adapting to new opportunities;
3 Ideological strategies: originate in shared beliefs and controlled through a collective vision;
4 Umbrella strategies: originate in constraints with leadership only in partial control;
5 Process strategies: originate where leadership controls process rather than content;
6 Unconnected strategies: originate in enclaves loosely linked within the organisation;
7 Consensus strategies: originate in consensus and implemented through mutual adjustment;
8 Imposed strategies: originate in the external environment and imposed on an organisation.

(Mintzberg 2007: 7–8)

Mintzberg produced these ideal types after studying a large number of organisations. They describe a certain 'pattern in a stream of decisions' within an organisation. For a scientific approach, his analysis of strategies within organisations is inadequate for two reasons. First, Mintzberg takes organisations (and their particular interests) as the starting point for his research. Mintzberg may argue that this is legitimate as each organisation needs to work out strategies for achieving its goal. However, his discussion of strategy is limited to this particular context and their interests in strategy.

Second, his research documents different types of strategies within organisations. However, it fails to distinguish between what a strategy is and how it is used by organisations. Despite his occasional reference to 'theory', he remains locked within the horizon of everyday common-sense living. Mintzberg's ideal types reflect the way in which strategy is used within an organisation and each use reflects some dominant interest within the organisation, whether the central leadership or a range of managers and workers. In short, what Mintzberg is documenting are strategies as distorted by ideology. The question that Mintzberg does not address is: what is a strategy? A theory of strategy would grasp the key significant elements and their relationships which constitute strategic planning. Mintzberg has little control over his investigation: his discussion of strategy variously includes policy as a stance, position or what is intended; it also includes application (or communication) in a particular time and place. A theory of strategy will distinguish strategic planning from both policy and application in a particular time and place.

As discussed in the previous section, the role of the functional specialty Policies is to work out the best direction for the future, the value or set of values that will become the vector for the future development of housing (and the totality of its purposes). Just as the functional specialty Policies envisages an understanding of policy within a theoretical context, so too does the functional specialty Systematics. Previously I proposed a theory of housing as a complete ordered hierarchical set of sets of related variable elements. Further, that the actual housing system in each country is an operative ordered set of sets of related actual elements.

This complex of systems, what we might call 'an ecology of systems', forms the context within which some new direction or vector of development is being proposed. What confronts the strategic planner is (i) an already existing ecology of systems, an ecology that has been found wanting, an ecology in which the current vector of its development has served the interests and self-aggrandisement of particular groups with consequent distortion of that ecology and (ii) a proposed shift in the vector of development envisaged in Policies. The task here for the functional specialty Systematics is to take this new direction or new vector of development and envisage courses of action for introducing new possibilities into this totality or ecology of systems, courses of action that will embody the shift in direction or vector of development.

Mintzberg, among others, proposes a view of strategic planning where a particular group – whether government, business, political party or community organisation – can get changes in their own interests when confronted by a situation in which a range of different players are seeking their own more or less competing interests. In contrast, the functional specialty Systematics is concerned with the integration of Policies into a larger systematic whole, one in which housing plays a role in an ongoing developing society. Once a new direction is discerned, a series of further decisions follows. The implementation of this new direction depends upon creative insights into the complete ordered set of sets of related actual elements that constitute housing. The current actual housing system expresses the interests or self-aggrandisement of particular groups. So, it is not only a matter of grasping a change within one or more sets of this ordered set but also of grasping how this change expresses the new vector of development formulated in Policies.

The strategic question which Systematics seeks to answer is: what courses of action will integrate a new vector or direction within the complex series of contexts that constitute an actual operative housing system? So, the functional specialty Systematics shifts away from a focus on how different groups relate to one another. Its focus is on how a proposed housing policy direction can be implemented within a range of contexts or systems (technological, economic, social, political and cultural) that constitute a society.

Communications and the practical question

In the mode of seeking an understanding of the past, the role of the functional specialty Research was to answer an empirical question and to provide the relevant material for answering a what-is-it-question. In the mode of implementing something new, the role of the functional specialty Communications is to identify or assemble together those events whose occurrence in this time and place will realise a course of action.

When speaking of Communications, I am not only referring to speaking and writing but to the whole range of expressions to which Research turned to gather data. As noted in Chapter 4, these encompass activities (both external and internal to each person), attitudes, beliefs, opinions, feelings, emotions, values, habits, expectations, motivations, skills, capacities, personal and social characteristics, material characteristics (of buildings, of habitats, of environments), language, clothes, decorations, art, music, sounds, dance, performance, video, film, symbols, signs, customs etc.

Like Research, Communications is concerned with the particular. However, it seeks to introduce some new particular expression. It is a return from theory to practice, from a complete ordered set of sets of related variable elements to an actual operative set of sets of related particular elements or particular practices that give effect to the course of action or strategy developed in Systematics. It points to the particular changes that housing practitioners, housing organisations or other related organisations need to implement to realise a course of action within the context of a new direction or new vector of development. This course of action may require a change not just in one person or, indeed, one housing organisation but in a range of persons in a range of organisations. These are changes in practices that require new or adapted skills in individuals (for example, the new skills required as a SHO introduces a new allocations system or rent-setting system, or more sophisticated asset management), changes in the roles that housing organisations play (for example, as one organisation takes up housing development while another specialises in housing management) or changes in the roles of other organisations (for example, as capital finance for housing is sourced from a range of financial institutions in addition to banks). Communications works with persons and organisations to change their practices, as implementation depends upon their co-operation.

The functional specialty Communications seeks an answer to the question: what practices/activities in this time and place will achieve a strategic course of action? It shifts away from implementation as a top-down command process in which those in authority direct others to change their practices. It focuses on the complexity of activities and working with housing practitioners and others to make particular changes in their relations with others, in institutions, roles and practices.

Implications for housing research

A key characteristic of housing research is its twofold nature where: research identifies, analyses and articulates a problem; and policy proposes solutions to these problems. It is these two broad genres that tend to dominate. This chapter has explored the second of these and my discussion has revealed some particular weaknesses. The *primary weakness* stems from the horizon of these researchers – they operate within the horizon of everyday common-sense living. This has three inter-related consequences.

The *first consequence* is the difficulty which housing researchers have in linking research and policy and thus drawing upon an adequate understanding of the area in which they are proposing policies. They have little idea of what theory is and its role as a heuristic within implementation.

The *second consequence* is that housing researchers shift immediately from 'understanding' a housing problem to proposing a solution. This understanding is an understanding of a particular purpose that housing has, rather than what housing is and its totality of purposes (as illustrated in Figure 5.1). The solution assumes some group agreement about this purpose, an agreement which may be odds with the interests of another group. So policies do emerge as new directions, but as new directions that pit a previous purpose against a new one, and thus we continue the ongoing conflicts about housing as one group dominates another as it replaces one purpose with another. The horizon of everyday common-sense living limits the scope of policies to those which maintain or enhance the interests of a particular group.

By operating within the horizon of everyday common-sense living, housing researchers are not developing policies founded on an understanding of housing, a theory of housing. They do not distinguish between what housing is and what its purpose is. As a consequence, policy-makers envisage housing policies from the perspective of a particular organisation (with its particular ideology linked to its interests in one particular purpose) rather than envisaging new directions for the totality of housing; strategic planners envisage strategic planning with its focus on contexts as a matter of working out how and to what extent this ideological stance can be imposed upon other groups; and practical implementation simply becomes the top-down application of policies. It is only when we can relate what housing is to its various purposes that we can envisage policies that integrate the totality of these various purposes, not just one isolated element.

The *third consequence* is their lack of precision regarding what they are doing. So questions of research methods become mixed up with the critique of other positions, questions of policy (looking forward) become mixed up with understanding the past; the differences between policy, strategic planning and communications are glossed over with some researchers focusing on some parts but adding in little bits of others.

A *secondary weakness* relates to those researchers who have some sense of foundations as a starting point for research. Each of the positions – social constructionist, critical realist or radical empiricist – outlined in this genre of research take their stand on arbitrary axioms, self-evident truths or first principles. Each can be evaluated by comparing it with the actual performance of the researcher adopting that position. Within the horizon of subject as subject, this performance, whether the dynamic of the desire to understand (that gives rise to research and learning), the dynamic of the desire to create something worthwhile (that gives rise to implementation), the dynamic of sociality or of intersubjectivity (that gives rise to co-operation and collaboration in the achievement of goals) can be thematised. This thematisation reveals criteria for competently doing these things. Moreover, it grounds an ordered set of questions (the eight functional specialties) that can be differentiated and related, a set of questions which will form the foundations of scientific endeavours.

The implementation functional specialties propose a more precise understanding of what researchers are doing when they are seeking to implement something new. Indeed, as with the research functional specialties, they propose a series of paradigm shifts.

Where the functional specialty Dialectic seeks an answer to the evaluative/ critical question 'what is the best of the past?', the functional specialty *Foundations* seeks to answer the transformative/visionary question 'who will I be?' or 'who will we be?' or, more specifically, 'who will I be as a housing researcher?' It proposes a shift in the foundations of research *from* a horizon based on arbitrary axioms, self-evident truths or first principles *to* the ongoing decision of the researcher to operate within the dynamic context of a differentiated and ordered set of questions, the functional specialties and the thematisation of the subject as subject.

Where the functional specialty History asks the historical question 'what dynamic or vector has provoked ongoing change in an actual operating housing system?', the functional specialty *Policies* asks the same question but in anticipation of the future, the policy question, what new operative dynamic or vector will best promote the future development of the current actual operating housing system into future systems? It proposes a shift *from* a focus on government or organisational policy – policies whose range is limited to one or other purpose which serve the interests of that government or organisation – *to* policy as a direction in which the totality of housing purposes are better realised.

Where the functional specialty Interpretation asks a theoretical question 'what is it?' and anticipates a complete ordered hierarchical set of related variable elements in which the role of the 'what' is specified within a series of contexts (technological, economic, political, cultural and personal), the functional specialty *Systematics* seeks an answer to the strategic question 'what course of action will best integrate a new operative dynamic, vector or direction within the complex series of contexts (technological, economic, political, cultural and personal) that constitute an actual operative housing system?' It proposes a shift *from* a focus on how a government or organisation will use housing to achieve its particular purposes in the context of other competing purposes *to* the integration of a proposed housing policy direction within its technological, economic, social, political and cultural contexts.

Where the functional specialty Research asked the empirical question 'what events are occurring in this time and place and to what extent are these events associated?' and provides the material relevant to answering a what-question, the functional specialty *Communications* seeks an answer to the practical question 'what practices/activities in this time and place will achieve a strategic course of action which will realise a new actual operative housing system?' It proposes a shift *from* implementation as a top-down command process *to* working with individuals and groups to make particular changes in their relations with others, in institutions, roles and practices.

Notes

1   Recall that in Chapter 7 on the functional specialty Dialectic, I outlined the radical transformation of horizon as we shift from the horizon of everyday common-sense living to the horizon of theory and from the horizon of theory to the horizon of subject as subject. In addition to these radical transformations, there is the gradual expansion of each of these horizons with each new discovery.

2   This is the final report of the National Housing Strategy. Leading up to this final report seven issues papers, fifteen background papers, three discussion papers, twelve monographs and nine consultation reports were published (National Housing Strategy 1992: 145–148).

3   Publications by these organisations include National Shelter (2001, 2004, 2009), ACOSS (1992, 2008), VCOSS (1989, 1990, 1991a, 1991b, 1992, 1993a, 1993b, 2008), McNelis (1992, 1993, 1996), Bisset & Victorian Council of Churches (1991) and Ecumenical Housing (1997).

4   This theme – from central principle to practices – is also reflected in the implementation of their research-policy linkages project (see above). It is also reflected in how the authors structure their thinking in the Options Paper (see below).

5   Section 2.2 of the Discussion Paper (2004: 14–18) describes in some detail the different characteristics of each model. They are summarised in their Table 1.

6   See their Table 4.1 on p.41ff for the relationship between the approaches and the specific options.

7   This criticism is not specific to the work of Jones and Seelig. It is a criticism of most housing research and policy-making even my own.

8   See http: //en.wikipedia.org/wiki/Henry_Mintzberg.

Part III

Functional Collaboration: A unifying framework

Dialectic   Foundations

History   Policies

Interpretation   Systematics

Research   Communications

Functional Collaboration proposes a complete ordered set of eight inter-related questions about housing: an empirical question, a theoretical question, a historical question, an evaluative/critical question, a transformative/visionary question, a policy question, a strategic question and a practical question.

Each question anticipates a different type of answer and different type of method by which it is answered. Part II explored these different methods and presented a more precise understanding of each of the eight functional specialties.

My problem in Part II, as I noted in its introduction, is that to understand the parts (the functional specialties) requires an understanding of the whole (Functional Collaboration), but to understand the whole requires an understanding of the parts. As discussed in Chapter 5, an insight grasps the set of relationships between elements. In doing so, it grasps the whole and the parts together. So, the presentation of the eight functional specialties in Part II presupposed a grasp of Functional Collaboration as a whole. Indeed, as the reader may have noted, the presentation of each functional specialty was not without references to the other specialties.

Part III focuses on Functional Collaboration as a whole, as a unifying framework.

9   Functional Collaboration, science and progress

> As the labor of introspection proceeds, one stumbles on Hegel's insight that the full objectification of the human spirit is the history of the human race. It is in the sum of the products of common sense and common nonsense, of the sciences and the philosophies, of moralities and religions, of social orders and cultural achievements, that there is mediated, set before us in a mirror in which we can behold, the originating principle of human aspiration and human attainment and failure. Still, if that vast panorama is to be explored methodologically, there is the prior need of method; if method is not to be a mere technique arrived at by trial and error, we must first know its grounds; and its grounds reside not in words or statements, not in concepts or judgements, not in experiences or acts of understanding, but in the principles, at once generative, constitutive, and normative of the human spirit in act ...
>
> Bernard Lonergan (1965 (unpublished): 14)

In the first chapter of his book *Wholeness and the Implicate Order*, the quantum physicist (and philosopher) David Bohm laments the fragmentation of individuals, society and the academy:

> fragmentation is now very widespread, not only throughout society, but also in each individual; and this is leading to a kind of general confusion of the mind, which creates an endless series of problems and interferes with our clarity of perception so seriously as to prevent us from being able to solve most of them.
>
> Thus art, science, technology, and human work in general, are divided up into specialties, each considered to be separate in essence from the others. Becoming dissatisfied with this state of affairs, men have set up further interdisciplinary subjects, which were intended to unite these specialties, but these new subjects have ultimately served mainly to add further separate fragments. Then, society as a whole has developed in such a way that it is broken up into separate nations and different religious, political, economic, racial groups, etc. Man's natural environment has correspondingly been seen as an aggregate of separately existent parts, to be exploited by different groups of people. Similarly, each individual human being has been

fragmented into a large number of separate and conflicting compartments, according to his different desires, aims, ambitions, loyalties, psychological characteristics, etc., to such an extent that it is generally accepted that some degree of neurosis is inevitable, while many individuals going beyond the 'normal' limits of fragmentation are classified as paranoid, schizoid, psychotic, etc.

(Bohm 2002: 1–2)

Bohm then goes on to discuss this problem of fragmentation and wholeness, its implications and the difficulties in finding a solution:

The question of fragmentation and wholeness is a subtle and difficult one, more subtle and difficult than those which lead to fundamentally new discoveries in science. To ask how to end fragmentation and to expect an answer in a few minutes makes even less sense than to ask how to develop a theory as new as Einstein's was when he was working on it, and to expect to be told what to do in terms of some programme, expressed in terms of formulae or recipes.

(Bohm 2002: 22–23)

The problem of fragmentation and complexity is echoed in a 2004 issue of *Futures* dedicated to transdisciplinarity. In the face of ever-pressing key questions about 'the natural and human-made environment' and about the economy and society, transdisciplinarity is understood as one way in which to bring together the resources of various disciplines to tackle these questions. In their introduction, Roderick Lawrence and Carole Després (2004: 399) describe it as follows:

[T]ransdisciplinarity tackles complexity in science and it challenges knowledge fragmentation…It deals with research problems and organizations that are defined from complex and heterogeneous domains…Beyond complexity and heterogeneity, this mode of knowledge production is also characterised by its hybrid nature, non-linearity and reflexivity, transcending any academic disciplinary structure…[T]ransdisciplinary research accepts local contexts and uncertainty; it is a context-specific negotiation of knowledge…Transdisciplinarity implies intercommunicative action. Trans-disciplinary knowledge is the result of intersubjectivity…It is a research process that includes the practical reasoning of individuals with the constraining and affording nature of social, organisational and material contexts…For this reason, transdisciplinary research and practice require close and continuous collaboration during all phases of a research project… [T]ransdisciplinary research is often action-oriented…entails making linkages not only across disciplinary boundaries but also between theoretical development and professional practice…Transdisciplinary contributions frequently deal with real-world topics and generate knowledge that not only address societal problems but also contribute to their solution…One of its aims is to understand the actual world...and to bridge the gap between knowledge derived from research and decision-making processes in society.

It is this fragmentation and complexity which Functional Collaboration addresses. But rather than seeking common ground in the multiple frameworks of various disciplines, it seeks their common ground in the structure of the questions and the methods by which they are answered as they emerge from the subject as subject.

Part II explored each of the functional specialties separately and proposed some major shifts in the way housing research is conducted. It also left some important questions unanswered: 'how are the functional specialties inter-related?', 'how are they are a unity?', 'how is Functional Collaboration a scientific approach?', 'how is Functional Collaboration a 'framework for collaborative creativity'?' and 'how does Functional Collaboration bring about progress in housing?'

Functional Collaboration: the inter-relationship and unity of the functional specialties

The starting point for *Research* is the events that have occurred or are occurring now, the events of everyday living. The function of Research is to ask the empirical question, time-place questions such as: 'how often has this occurred?', 'how often is the occurrence of this associated with the occurrence of that?', 'who said what?', 'who felt what?' In housing research, we are most interested in the expression of ourselves and of our sociality or inter-subjectivity through the construction of our social world. This inter-subjectivity, anything that pertains to the togetherness of human beings (Lonergan [1972] 1990: 359), is expressed through an ever-changing manifold of external activities, internal activities, attitudes, beliefs, opinions, feelings, emotions, values, habits, expectations, motivations, skills, capacities, personal and social characteristics, material characteristics (of buildings, of habitats, of environments), languages, clothes, decorations, art, music, sounds, dance, performance, video, film, symbols, signs, customs etc. It is these expressions that become the data that is attended to by Research.

While Research attends to and gathers data, its orientation is towards making this data available to Interpretation. In Chapter 5, we noted that the higher value orders the particularity of the lower value. Interpretation does this in two ways. First, it provides the heuristic which guides the work of Research. This heuristic is the most up-to-date theory. But unlike History which uses this theory as a heuristic or tool for understanding what is happening in a particular situation, in a particular time and place, the focus of Research is on questioning and testing the current theory. By being attentive to the data of everyday living, researchers are alert to events not accounted for in the current theory, seeking out anomalies between the data and the theory. So, for instance, Research might discover data that suggests that one element in the theory does not belong here but belongs there, or does not sufficiently distinguish one aspect from another, or that the role of some higher value in ordering the particularity of a lower value is missing.

The view of Research as to what is anomalous between the current theory and the gathered data becomes most pointed in the way in which Research presents its data. A second way, then, in which Interpretation orders the particularity of Research is in the way data is presented to Interpretation. Research can make data available to Interpretation in any number of ways: as descriptions of events; as narratives; as a collation of quotes or as selected quotes from those interviewed; as selected audio or video recordings; as raw numbers or frequencies of answers to questionnaires; as tables containing numbers, frequencies, proportions, percentages, ratios, rates, regressions, standard deviations, maxima and minima etc.; as different types of graphs (ranging from simple line, column, bar, pie, area, scattergram graphs to more complex ones); as different types of maps or three-dimensional models etc.

In doing so, they are transforming one form of human expression into another. So, the materials of a population and their characteristics, the materials of the views, attitudes, feelings etc. of participants in interviews, the materials of a sequence of events, the materials of documents, recordings, performances, photos etc., all of which are expressions or products of human creativity, are gathered together in some way and transformed into another form of human expression, one which seeks to communicate some new and significant elements to Interpretation. This new form of expression becomes the materials upon which Interpretation operates. It can be provocative. It can be suggestive. Or, indeed, it can inhibit Interpretation. For example, '[i]t is easy enough to take the square root of 1764. It is another matter to take the square root of MDCCLXIV' (Lonergan [1957] 1992: 42; more generally Chapter 1).

In this way, Research has its specific function in the cyclical process of discovering new meanings in housing and creating new developments in housing.

We are not satisfied with just gathering manifold and disparate data on occurring events. We want to understand it, understand its meaning and significance. The function of *Interpretation*, then, is to answer the theoretical question: what are the elements and their relationships that constitute housing? Interpretation faces the challenge of taking the anomalous data provided by Research and, in some way, incorporating it (by adjusting the current theory or by developing a new theory) and so seeking a better answer to this question, one which takes account of the anomalous data. This answer is an explanatory definition or theory which selects from the manifold of data what is relevant, significant and important to constituting housing, to its occurrence.

As a heuristic, a theory of housing developed in Interpretation plays a role in other functional specialties. By answering the question 'what is housing?', it controls the scope of housing researchers' inquiries, limiting this scope to the complete ordered set of sets of related variable elements that constitute housing and its particularities. The particularity of Interpretation is thus ordered by these other functional specialties. For instance, History requires a heuristic of housing that is universal, that can be used in any place and any time. It requires a heuristic that is global, that is not limited to any country, society or culture (or even to a number of countries, societies or cultures), whether in Europe, Asia, the Americas,

Africa or the Pacific, whether current or in the recent past or in the remote past, and whatever the particularities of housing as they vary from country to country, from society to society and from culture to culture. It requires a heuristic that is transcultural. In addition, History requires an adequate heuristic, one which incorporates not just the set of related variable elements that constitute housing, but also a complete ordered set of sets of related variable elements that constitute the particularities of housing. History requires an omni-disciplinary heuristic, one which incorporates in a structured way the technological, economic, political and cultural aspects of housing. With such a heuristic, History can then identify sequences of actual operating housing systems.

We live not in a static world but an ever-changing dynamic world. Housing at one point in time is different from that at another point in time. Our understanding of housing at one point in time is different from that at another point in time. Both housing and our understanding of it are continually changing. But what brings about such change? The function of *History*, then, is to answer the historical question: what dynamic or vector has provoked ongoing change in an actual operating housing system? History takes as its data a theory of housing from Interpretation. This theory becomes a heuristic for identifying an actual operating housing system at different points in time. Theory provides us with a heuristic, a tool for understanding what is happening in any particular situation, in any particular time and place. While this heuristic defines housing as an object of inquiry, the actual operative housing system will be particular to the situation and differ in time and in place. Thus, theory provides us with a set of questions for understanding what is happening. As Paul St Amour (2009: 19), commenting on Lonergan's theory of the economy, puts it:

> Being able to think about the economy with the aid of such a theory complements tremendously the resources of common sense understanding, no matter how refined, prudential and morally informed those may be. It vastly broadens the horizon of discourse by making possible an array of questions which would not likely occur in the absence of the theoretic contribution.

These actual operative housing systems form a sequence, sometimes stable and changing slowly, sometimes unstable and changing rapidly, sometimes with only minor changes to one or other element of housing, sometimes with major changes in all elements. History seeks to grasp the vector within this sequence of actual operating housing systems, what it is that motivates the changes over time in the sequence of actual operating housing systems. The vector of change in housing can be any one or a combination of the values in which housing plays some role. It is these values which motivate the changes in the different elements of housing over time. Indeed, the vector of change becomes taken for granted within a society and, as a result, it changes slowly.

We are, however, not satisfied with simply grasping these vectors of change in housing. We also want to know whether they are worthwhile. The function of

Dialectic, then, is to search out the best of the past. Dialectic asks the evaluative/ critical question: what is the best of the past? It takes as its data the complex vectors of change grasped by History. These vectors are deeply embedded within the taken-for-granted everyday world of a particular culture. They operate not just within the field of housing but are vectors of change across many fields. The task of Dialectic is to evaluate this data, to work out whether the priority afforded the particularisation of this vector promotes the development of the whole of society or whether it gives predominance to one or other value as a vector of development at the expense of other values within the whole, and so 'promotes' the dominating power and self-aggrandisement of individuals and groups and their interests. The horizon within which Dialectic evaluates this vector of history is the horizon of the subject as subject. It is within this horizon that Dialectic grasps whether the particularisation of this value as a vector of development proceeds from the dynamic of the subject. The work of Dialectic results in an integral view of the current vector: an appreciation of what has been achieved through that vector; a critique of its limitations; and the location of the vector and its role within a larger whole.

Foundations thematises the subject as subject and so provides the grounds for evaluation. In this way, Foundations orders the particularity of Dialectic. Consequently, the results of Dialectic are never definitive. Foundations demands that Dialectic operates at 'the level of our times'. The horizon of Dialectic is always on the move, seeking the best of the past. So, it is no longer acceptable for housing researchers to operate within the horizon of everyday common-sense living; it is no longer acceptable for them to operate within some horizon of theory. By thematising the horizon of the subject as subject, Foundations provides a precise understanding of theory within the larger context of the process from understanding the current situation to implementing something new, it provides the grounds for ongoing development of the subject as subject, of technology, of economics, of politics and of culture.

Dialectic provides an evaluation of the complex vectors of the past, searching out the best of that past, the best that promotes development. It emerges with a discovery, a new sense of what promotes progress. It emerges with a new sense of priorities, a sense of changing directions and orientation. Dialectic does not altogether reject the previous vector of development. Rather, sensing its limitations, it locates this vector within a larger whole. It is within this sense that a new direction needs to be forged, one which requires a decision that will transform the future. It is a discovery which demands expression. Foundations seek to articulate the shift that this decision incorporates, a shift that will become the horizon for future development. For housing researchers, the discovery of Functional Collaboration as a scientific approach to housing is one such discovery. It is within the horizon of Functional Collaboration that housing researchers can progressively improve their capacity and skills to understand housing, offset the prejudices and biases of the dominant culture and provide practical advice to decision-makers.

So, a decision by housing researchers regarding Foundations provides the horizon for the ongoing development of housing. It provides the grounds for something new, a new direction. *Policies* articulates a new vector or new basis for future development. History grasped the vector of past change, and Dialectic evaluated whether this vector brought about ongoing development or whether it brought about stagnation and destruction. Many different vectors could be introduced, each of which is in the interests of a particular group. Policies, however, grasps the vector that will bring about ongoing future development of the whole. It will promote housing within the totality of purposes it plays.

For this vector to be practical it must, however, operate within broader technological, economic, political and cultural contexts. *Systematics* considers how this new vector of development can operate within the ecology of systems that constitute these contexts. Systematics is concerned with working out a strategic course of action among many possible ones through which this new vector could be implemented. It is not simply a matter of any course of action but the best course of action which integrates with and emerges from the possibilities inherent within the current context, the manifold of systems that constitute a society.

Communications takes this best possible course of action and works out the time-space details for implementing it. Communications provides practical advice to decision-makers, here and now, as to what new practices will realise this course of action: what changes in understanding will be required; what adaptations to current skills and what new skills individuals will need to develop for these new practices; and how institutions can be adapted to these new practices.

Communications provides practical advice to decision-makers. This advice may or may not be accepted. It may or may not be fully implemented. It may or may not be adequate to the task. Some new housing situation will emerge from what was implemented. This new situation provides the data for Research.

In this way, the functional specialties can not only be distinguished one from the other but they can be understood as inter-related: the results of one functional specialty become the data which the next functional specialty operates on. Each adds something to the results of the preceding functional specialty as we move forward from one housing situation to the next. The unity of the functional specialties is the unity of a process, the process through which the subject as subject brings about something worthwhile, the process which takes the subject beyond the current situation into a new one and the process whereby a group of people, a society, addresses their current problems by creating something better.

Functional Collaboration as a scientific approach

At the beginning of Chapter 3 I asked: what is a scientific approach to any subject? What distinguishes a scientific approach from any other approach? Can we have a scientific approach to social science? At that point I referred to Lindberg's review of the meanings of science (1992). I noted that the term 'science' is ususally reserved for the natural and formal sciences and proposed that science is not simply an advancement of knowledge but also an advancement of better living. Our question here is whether Functional Collaboration is a scientific approach.

To reach a contemporary understanding of science is no simple matter, however. In seeking an answer to the question 'what is science', we might proceed by examining the work of scientists in various fields or the work of experts in the area (philosophers of science) and adopting one or other approach. But to do so plunges us back into many of the issues we raised in Chapter 5 and Chapter 7.

We can define science descriptively or we can define it explanatorily. A descriptive definition of science will no doubt appeal to science as it is practised, particularly as it practised in the natural sciences. So, we might characterise or define science in terms of the reliability of its methods for collecting data, its rigour, precision, objectivity and ability to predict future events (Lindberg 1992; Flyvbjerg 2001). My previous discussion of descriptive definitions, however, finds this approach inadequate. A descriptive definition highlights associations within particular sciences. It has yet to distinguish between science and its characteristics. It has yet to distinguish a complete set of elements and their systematic relations that constitute science (and distinguish science from other things). It has yet to locate science within contexts that explain its particular characteristics in different sciences.

An adequate answer to the question 'what is science' will be an explanatory definition. We begin with some understanding of science. We inquire into the practices of a whole range of sciences. We mess with the data trying to interpret it, trying to make sense of it, trying to distinguish science from its many purposes (including the promotion of viewpoints on science which aggrandise particular sciences). We use 'fantasy and lateral thinking' as we reach for a discovery, until we eventually grasp through insight the complete set of related variable elements that constitute science. It is a theory of science. It will distinguish between science and its various purposes. It will distinguish between science and its particular role or function in the formal, natural, biological, psychological and social sciences. It will play a role in the higher viewpoint of human living.[1]

Throughout its history, the understanding of science has been a developing one. Karl Jaspers argues that the period between 800 to 200 BCE was a pivotal time in the history of humankind as 'man everywhere became aware of being as a whole, of himself and his limits' – across China, India, Iran, Palestine and Greece (1954: 100; see also Jaspers 1953; Lonergan [1972] 1990: 85–99, 257–262; McShane 1984; Drage 2003). Bruno Snell describes the discovery of the mind by Greeks beginning with their literature and culminating in the thought of Socrates, Plato and Aristotle (1953; see also Lonergan 1972). 'With Socrates there explicitly emerges the theoretic 'what question.' Plato established the reality of the idea or form. Aristotle empirically links the 'what question' and the 'form' by grasping the pivotal function of 'insight into phantasm'' (Shute 2013a). Aristotle's *Organon* was the standard for scientific method for centuries until Galileo, the father of modern science, developed new scientific procedures in the sixteenth century. Francis Bacon's *Novum Organum* ([1620] 2005) replaced Aristotle's *Organon*. In the nineteenth century, John Stuart Mill's *A*

System of Logic ([1882] 2009) replaced Bacon as the standard. Quantum mechanics in the early twentieth century raised foundational questions about traditional classical science and founded itself on statistical method. Giambattista Vico in his *Scienza Nuova* ([1744] 1948) highlighted the importance of historical method. Auguste Comte, Karl Marx, Herbert Spencer, Friedrich Nietzsche, Georg Simmel, Emile Durkheim, Max Weber, Wilhelm Dilthey, George Hebert Mead, Talcott Parsons, Robert Merton, C. Wright Mills, Karl Popper, Hans-Georg Gadamer, Alfred Schutz, Thomas Kuhn, Jurgen Habermas, Paul Feyerabend, Anthony Giddens, Michel Foucault, Pierre Bourdieu and Roy Bhaskar, among many others, have each made their contribution to method in the human and social sciences.

Throughout history, science has always sought to work on the basis of the most developed foundations for research. What was once an acceptable basis for science no longer is. Each development highlights some aspect of science: logic and theory; empirical evidence; radical changes in scientific understandings throughout history; that science operates and develops within communities; a critique of these communities and their presuppositions and the recognition of the limitations of their scientific horizons. Nearly thirty years ago, Matthew Lamb (1985: 76) summed up the current position on science as follows:

> Within the extensive debates and disagreements among philosophers of science, there is an emerging consensus about the illusory character of this underlying presupposition of a logically pure objectivism in natural science… The question is no longer how to modify the scientistic objectivism of positivism or logical empiricism, but what paradigms will eventually take their place…Nevertheless, there seem to be two generally accepted judgments on the direction of such developments: (1) rationality can no longer be defined solely with reference to the procedures of mathematics, logics, or the natural sciences; and (2) even in these domains – and a fortiori in others – attention is shifting from theories of theories to the heuristic performance or praxis of theorizing. The deductivist ideals of the coherent and complete criteria for rationality provided by theory qua theory are gone, even in mathematics. Theories as formally logical systems are, to paraphrase Kurt Godel, either incomplete and consistent or complete and inconsistent. Rationality cannot be identified with ideals appealing to non-existently and impossibly complete and consistent foundations in theory qua theory. There is a shift from scientistic objectivism to the questioning procedures or praxis of communities of inquirers.

The question 'what is science' is not a simple one. It is an ongoing question: what is now an acceptable basis for a social science? This is a question for the functional specialty Dialectics which would recycle, evaluate and critique past understandings of science seeking out the best of these understandings and working towards an understanding that integrates the various aspects of science noted above. Such a task is beyond this book.

Is Functional Collaboration a scientific approach to housing? Reaching an answer to this question is not simply one of developing a theory of science and then comparing our answer with our understanding of Functional Collaboration. In proposing Functional Collaboration as a scientific approach to housing, I am proposing a new understanding of science. This new understanding (and doing) of science is based upon the praxis of communities of inter-related inquirers. It brings together within one framework past discoveries of new scientific methods. As Michael Oxley has noted in relation to comparative housing research, methods and purpose go together. Different aims demand different approaches and each method has to be fit for purpose (Oxley 2001: 103; see Chapter 3 above). Put simply, science has its origins in the mind of the researcher seeking answers to questions about something that puzzles them. Functional Collaboration, then, may be understood as a scientific approach to housing which draws the significant, relevant and essential methods that constitute science. Different sciences particularise these methods to fit their particular purposes and currently in the course of their development different sciences have highlighted, indeed have been pre-occupied, with one or other method to the exclusion of others. So the natural sciences have focused particularly on the relationship between empirical evidence and theory. The social sciences have focused on history, critique and implementation. Functional Collaboration brings all these methods together into one framework.

I began my discussion of the functional specialties with Research. But the real starting point for a scientific approach to housing is Foundations. Prior to any housing research, Foundations sets the context for a scientific approach. It thematises the set of eight questions, the type of answers each anticipates and the method by which each question is answered. It grounds the different functional specialties and different types of housing research. In this way, Functional Collaboration envisages an explanatory definition of science, one founded in the subject as subject and the decision of the housing researcher to operate in accord with the dynamic of understanding. Further, it provides the horizon within which sciences operating in different fields, whether formal, natural, biological, psychological, social sciences etc. and the different disciplines within them, can be implemented through vectors or directions that promote the future development of the whole (Policies), through courses of action that integrate these vectors within a complex series of contexts (Systematics) and through activities that are practical for each particular science (Communications).

Functional Collaboration as a framework for collaborative creativity

Collaboration between researchers is seen as the way of the future (Adams 2012, 2013). This may be a new discovery for researchers. Yet, it has always been the case, even before economist and moral philosopher Adam Smith observed that a division of labour 'in so far as it can be introduced, occasions, in every art, a proportionate increase in the productive power of labour' ([1776] 2009: Book 1, Chapter 1, Locations 134–143). Collaboration is constitutive of society.

Our actions are virtually always contributions to a wider communal or social project involving others. Action is seldom action in isolation. Most usually our acting is co-operating. Our initiatives are contributions to joint projects, tuned to fit in with the contributions of others, fashioned as links within wider chains of actions which bring societal projects from inception to implementation.

(Melchin 1994: 23)

Within the horizon of the subject as subject, the current forms of collaboration are the ordering of our desire to understand by our desire to create something worthwhile and our desire for sociality.

Significant research is never the result of the work of one housing researcher. It is more commonly the result of researchers working together. Even researchers working alone draw upon the work of others through literature reviews, conferences, institutions, etc. The work of each housing researcher is part of a vast array of inter-linked and related activities involving many different researchers.

Collaboration, however, has become a problem in the context of complexity. Traditionally we have resolved the problem of complexity through specialisation, either field specialisation or subject (discipline) specialisation (see Chapter 2). As specialisation has continued apace, fields and subjects divide into sub-fields/subjects and into sub-sub-fields/subjects. While this solution deals with increased complexity of fields/subjects, it has also served to fragment the academy and society and widen the gap between theory and practice, between research and policy and between local and global needs and demands. As a result, it becomes more difficult to grasp the whole, to relate disparate fields and subjects to one another and within fields and subjects to relate disparate methods to one another. Collaboration within specialties is possible where researchers operate on common presuppositions with common or complimentary approaches or methods. This is limited and there now emerges a demand for a new way of dealing with complexity which will more effectively promote collaboration among researchers such that they can find solutions to our pressing social problems.

Functional Collaboration is a response to this demand. It proposes a new form of specialisation and collaboration, one based upon functions within the process of moving from the data of the current situation to results that provide practical advice to decision-makers. This form of collaboration does not depend upon personal or social relationships or ad hoc relations between institutions. Rather, it depends upon how one type of work relates to another type of work within the process from data to results. As functional, it is global in scope. It envisages all housing researchers operating within functional specialties, wherever they are. In this way, it is oriented to praxis; it links research, policy and implementation; it links theory and practice; it cycles through the process from data to results producing cumulative, ongoing and progressive results.

As outlined in Chapter 5, the role of the functional specialty Interpretation envisages a developing heuristic which relates and integrates all sciences and disciplines; a theory of housing envisages a complete context of housing in which the full range of purposes are related to one another. It is a heuristic which holds the whole together. It is a heuristic which provides a control throughout the process. Where different sciences and disciplines break up this whole and consider the future in terms of their component, Functional Collaboration through the implementation functional specialties considers the future of the whole and each of its components in relation to this whole.

An individual researcher can differentiate the eight inter-related questions that constitute a scientific approach to housing. The effectiveness and productivity of housing research, however, can be enhanced if many or all housing researchers operate within the framework of Functional Collaboration. They can understand how their contribution relates to that of other contributions; they can more effectively draw on contributions relevant to their work and pass on their results to others. The more the functional specialties develop, the more they will develop and refine their methods. The more housing researchers understand and appreciate their reciprocal dependence, their function within the whole research process, the more expert they will become at passing on their results to others and the more effectively housing research will deliver sound practical advice to decision-makers (Lonergan ([1968] 2010: 462).

Functional Collaboration is a 'framework for collaborative creativity' (Lonergan [1972] 1990: xi) that, on the one hand, maintains the unity of a scientific approach to housing and more effectively brings about better human living and, on the other hand, promotes creativity and diversity within housing research.

Making progress in housing

In Chapter 1, I raised the vexed issue of progress. I noted that decision-making is central to the process of change. While different groups advocate different and conflicting courses of action as the best, their underlying intent is to bring about something better, some improvement in housing or some related purpose. The question we face is whether these decisions have brought about progress rather than deterioration. Who is to say whether something is progress? By what criteria? If progress is not simply a matter of who has the power to effect changes in their own interests, how can we get beyond particular views on progress within a particular culture with its bias towards the interests of dominant groups? How do we get beyond the stupidities, greed, entrenched interests and corruption of decision-makers as well as their unacknowledged presuppositions that maintain an unjust status quo?

When we think of making progress in housing, we may think of improvements over the past century, improvements in housing standards and quality, in housing management, in housing finance, in housing design, materials, technologies and building techniques, in the alleviation of housing-related poverty, in tenant and consumer rights, in the separation of residential areas from industrial areas, particularly noxious and malodourous industries, in the development of utilities such as water, electricity, gas, sewage and storm water etc.

An improvement is not just a good idea but something that is actually implemented. As an improvement, it is practical. It is not just a once-off, ad hoc or random occurrence of something new but rather a regularly recurring one. It is the result of a set of regularly recurring activities that have become a habit or institutionalised. If it is to be practical, however, it must take account of the context within which it is implemented – it must be strategic. Too many good ideas have failed because they were neither strategic nor practical. The idea is a good one insofar as introduces into history something new that is worthwhile and it expresses an aspiration for the future. As a good idea it presupposes some discovery: some grasp of what is happening and its significance; some grasp of the vector that is currently operative in the current situation; some critique/evaluation that shows up the limitation of the current situation. An improvement, then, has its conditions.

Here I am pointing not to the actual improvement but rather to the process that constitutes it as an improvement. If we are to make progress in housing, then it is vitally important we understand how improvements come about. We then have criteria for distinguishing improvements from setbacks; improvements have been subject to the whole process whereas setbacks slide over aspects of this process; ignore understandings or critiques of the current situation, or ignore the strategic context or practicalities.

The different elements in this process are those I have been outlining as Functional Collaboration whose end result is practical advice in local situations. I would suggest then, that Functional Collaboration is a central element in a theory of progress. It is a discovery that progress in housing, indeed in any area of human endeavour, is brought about or constituted by asking and finding new answers to eight inter-related questions.

Functional Collaboration is not an idealist fantasy, a dream of some future utopia. Rather, it formulates a discovery of something that already operates, albeit confusedly. It articulates how human history brings about progress. It articulates the elements, their structure and their relationship to one another. It is a discovery that anticipates opening up a vast range of possibilities in any area of human endeavour as we make a conscious decision to put this process in place.

This discovery will not put an end to disputes about the best way forward. It is a discovery that shifts the probabilities of making progress in housing, shifts these probabilities from *ad hoc* and random to more consistent, ongoing and cumulative. Indeed, as the complete set of inter-related questions to which we need new answers to move forward, this discovery pinpoints more clearly the source of disputes (whether they relate to one or other question), pinpoints the blocks to moving forward. In this way, it points to the work that needs to be done to resolve disputes. Functional Collaboration provides a framework for collaborative creativity for facing fundamental differences and disputes and working towards their resolution.

Yes, different groups propose different ways of moving forward with some more deliberately advancing views that enhance/entrench their own interests. It is the role of Dialectic to work out the best of the past, the dynamic of history that has brought about progress in housing or some other area of human endeavour. Dialectic reveals, appreciates, evaluates and critiques the dynamics of history including the interests of dominant groups seeking their aggrandisement. The criteria for progress, however, lie within the dynamics of the subject as subject: the dynamic of the desire to understand; the dynamic of creating something worthwhile, the dynamic of sociality, the dynamic of being-in-love. In themselves these dynamics are open, they do not point to any particular understanding or product as progress. There are no definitive external or objective criteria for progress. There are no *a priori* answers as to what is best. We learn what promotes progress by 'retracing our steps', by reflecting upon, evaluating and critiquing our history as individuals, as societies, as cultures and as humanity in solidarity with one another, our natural environment and the cosmos. We have to work this out in the great experiment of history, learning as we go.

So, if housing researchers and policy-makers are to make progress in housing, it is essential that each reveal their grounds for progress. 'So, [I] have to come up with those very revealing last chapters: Chapter Y: this is what I think is progress, and Chapter Z: these are the grounds upon which I would fantasize the future' (McShane 2004c: 6.3, p.7). History reveals the flow of changes in the past and what drives the flow of change. However, History does not distinguish between what is progress and deterioration in this past. As a pointer to future directions, housing researchers need to work this out. As I noted above in Chapter 7, what is required of housing researchers is that they search out and retain the best of the past (that which is the product of the pursuit of answers to questions) and discard what is shown to be inadequate, irrelevant and destructive (that which is the product of the interference of the dynamic of the desire to know with other dynamics). Progress is not a logical process nor is it fixed for all time. It is ever on the move as the future builds upon the discoveries of the past. The horizon of the subject as subject provides a heuristic of progress. At any time, the future is forged from the best we can come up with about the past. So, to promote progress we need to live in view of the best of the past.

Progress in housing is no easy matter. Sustained progress in housing is even more difficult. The fragmentation of housing research is a barrier to progress. In the long term, Functional Collaboration will be the very slowly evolving solution to the problem of making progress in housing.

Note

1   See Chapter 7 at the end of the section 'What is theory?', as well as the section 'Dialectic and housing researchers'.

Conclusion

How can we create a housing system worth inheriting? How can we pass on to future generations a housing system which not only meets the minimal needs of households for shelter but one in which, in the words of Peter King (2008), we can 'dwell': where there is 'privacy and security', where there is 'the possibility of intimacy', where 'intimacy can be protected', where 'we can be complacent – take what we are and what we have for granted – where we can 'exercise some control' and make choices, where we can express our self and where we share 'a common interest based on the shared experience of housing itself'?

It is this question that permeates my considerations of the role of housing research in the development of housing. It is this question which confronts authentic and conscientious decision-makers as they seek to grasp the possibilities of bringing about something new in history.

At the heart of research, whether in relation to technology, to the economy, to social or political institutions, to culture, there is an aspiration to transform the world in which we live, to make our world better, to improve our standard of living, to improve the quality of our lives. Within this aspiration lies the hope of a better future. But for this aspiration to flower it needs nurturing. In the disillusioned academic, in the mind-numbing routines of bureaucracies, in the failures of hopes and dreams, it can be destroyed and nullified. In the self-aggrandisement of individuals and groups, it can be perverted and narrowed. In the lack of strategy and practical understanding, it can be naively optimistic. Most housing research is directed towards something better than the current situation. It may be directed at a small improvement in our understanding, a small improvement in one aspect of housing. It may envisage much grander transformations. It is this aspiration for something better that Jones and Seelig (2004) point to in their discussion of champions and reformers for improving research-policy linkages.

In the aspiration to transform housing, I began by identifying two sets of problems. The first set regards practical advice to decision-makers within the complexity of the totality of past and present events, processes and practices that make housing what it is today. This totality is continually changing, continually calling for decisions that meet the demands of the current situation. Moreover, decision-makers are faced with a disparate array of housing research. It operates across many disciplines. It is characterised by different epistemological and ontological approaches and a broad range of methods. Its purposes range from theoretical to strategic to practical. Each researcher presents a different perspective, has something significant to say and would hope to have some impact on housing outcomes. Housing research is very fragmented. Housing researchers have very little sense of how different types of research relate to one another. Housing research has largely failed to provide practical and relevant advice to decision-makers considering the future.

The second set of problems regards the limitations of the personal and communal horizons of both housing researchers and decision-makers. These horizons are dominated by everyday immediate concerns. Whether deliberately or unconsciously, housing research and decisions about the future of housing are apt to support the interests and self-aggrandisement of dominant groups. As a result, the development of housing is skewed, narrowed and resisted. The aspirations and hopes of individuals and groups for something better remain unrealised.

As the discussion has developed, however, not two sets but eight sets of problems have been identified. They regard empirical evidence, the significance of housing, the vectors of history, evaluation/critique of both decision-makers and housing researchers, the foundations for moving forward, envisaging new directions, strategising these new directions and working out changes in practices.

If we are to move forward we need to address each set of problems. If we are to understand 'the vast panorama' of past housing events, processes and practices and look to the future, 'there is the prior need of method' (see the epigraph at the beginning of Chapter 9). If we are to find solutions to our pressing housing problems, we need a framework which will address each set of problems, hold the diversity of housing methods together and provide practical advice to decision-makers. The development of such an integrating framework is not easy, as José Ortega y Gasset ([1946] 1998: 72) notes in his *Mission of the University*:

> The need to create sound syntheses and systematizations of knowledge, to be taught in the 'Faculty of Culture', will call out a kind of scientific genius which hitherto has existed only as an aberration: the genius for integration. Of necessity this means specialization, as all creative effort inevitably does; but this time, the man will be specialized in the construction of the whole.

One way of dealing with this set of problems is by dividing up the data. Another is to divide up the results according to disciplines. While both these divisions of labour or specialisations allow researchers to deal with limited areas of housing, they have two limitations. First, these specialisations selectively deal with only some aspect of housing, and so, they break up the unity of housing. Second, it is unclear how the work of one researcher dealing with one aspect relates to the work of another.

In the mid-1960s Bernard Lonergan proposed a third form of specialisation, Functional Collaboration. It divides up the work according to stages or functions within the whole process from data to results. With this form of specialisation, researchers deal with housing as a totality, maintain the integrity of the data (deal with it as a whole), develop a more comprehensive or complete theory of housing (one which encompasses all the disciplines), grasp the dynamic operative within the history of housing as a whole, locate housing as whole within the broader context of a society, propose and implement policies that meet not only the demands of one interest group but rather housing as whole. The whole process from data to results is divided into distinct functions so that the work of one researcher is related to the work of others, with each drawing upon the work of researchers in one function and contributing to the work of those in another.

Within this form of specialisation, we begin with an already operating housing system, a system which we seek to improve, transform or, indeed, go beyond. The goal of Functional Collaboration is to meet the sets of complex emerging problems head on by expanding our understanding of these social constructions, bringing them under some control and increasing the possibilities of transforming them for the better. By dividing up the process of housing research, we can discern a complete set of inter-related questions, each of which addresses one of the sets of problems. So, this form of specialisation begins with the functional specialty Research which asks the empirical question: what events, processes and practices are actually occurring? It ends in the functional specialty Communications which asks the practical question: what practices will transform housing?

How are we to reach an understanding of this form of specialisation? How are we to distinguish the different stages in the process that takes us from the current situation to practices for its transformation? My starting point was a phenomenology of research in which I appealed to the experience of researchers.

> One way of leading the scientist, or the common-sense person, to some appreciation of the existence and nature of this science of methodology is to indicate the various problems of method that are present in contemporary culture. The difficulty of such indications, however, is that they can be appreciated only by the person engaged in the relevant field. Thus, there are basic methodological problems in contemporary physics, particularly in quantum mechanics and relativity, but for any reader who is not actually engaged in these fields the existence of such problems will be a matter of belief...One may say that a problem of method is a problem of know-how or know-what-we're-doing. Putting the problem this way helps to throw the stress where it should be: on the person doing the science and not on the science conceived of in some strange way as independent of mind. Now from this angle the problem in quantum theory would be expressed as 'What-are-we-at in doing quantum mechanics?'

> (McShane 1974: 4)

Or, 'what-are-we-at' in researching or theorising or understanding history or evaluating or creating housing?

A phenomenology of research asks the researcher to heighten their experience of asking and answering questions, to heighten their experience of doing research.

It asks: what are you doing when you are doing research? What is intended or achieved by each of these activities? How do the various activities relate together? What do combinations of these activities intend? It is only insofar as a researcher has come to appropriate what they are doing when they are doing research that they will be able to distinguish different types of questions and the respective operations whereby they go from not understanding something to an expression of some new understanding; and that they will be able to identify the sequence of questions: before we answer a what-to-do-question, we need to answer an is-it-worthwhile-question; before we answer an is-it-worthwhile question, we need to answer a what-is-going-forward-question; before we answer a what-is-going-forward-question, we need to answer a what-question.

In doing this, researchers will shift away from one particular understanding of science as somehow independent of the mind of the researcher and point squarely at themselves as the source of science. Moreover, they will recognise that their understanding depends upon the competence with which they perform certain activities; they will be able to identify the process whereby their understanding develops, their personality develops and their living develops; they will be able to grasp how to make progress as a researcher; they will grasp that many of the problems of science emerge because a researcher does not understand what they are doing when doing research. It is through this experience that a researcher begins to distinguish between the procedures of everyday common-sense living and the procedures of science. Where in everyday common-sense living we ask all sorts of questions in many different ways, in science we are more precise about our questions, about the ordering of and relationship between our questions, about the answers anticipated by those questions and about the method whereby we answer each question.

Part II highlighted different genres of housing research. Each has a general orientation around a particular question though this remains undefined. Moreover, authors do not consistently operate within one orientation but continually shift orientation. These shifts point to the largely everyday common-sense horizon of most housing research. Housing researchers are socialised and acculturated within some culture before they become researchers. They learn the skills necessary for living: how to deal with any situation that arises; what is important and what is not; that some things are more important than others. In doing so, they learn what interests others and how to use those interests, they learn how to exert their power and influence etc. Housing researchers are formed by the same or by a similar culture that values and produces housing. As they move into research, they not only learn the skills for their particular profession, they also bring the skills, the values, the beliefs, the interests, the prejudices and biases of the culture in which they are immersed.

As different cultures express different values, beliefs, interests, prejudices and biases, so too do researchers understand housing differently and propose to transform it in different ways. Were it simply a matter of understanding housing, making proposals for its transformation and writing about it – simply a dialogue among professionals – such differences would be of little moment except to the protagonists. But more is at stake, for different understandings and different proposals have their differing results in housing practices, and these practices may be disastrous for some people while being beneficial to others, may destroy some lives while enhancing others. So not only do questions emerge about housing practices, they also emerge about researchers themselves, about how their skills, their values, their beliefs, their interests, their prejudices and their biases distort their understanding of housing and their proposals for its transformation. Some researchers can deliberately make choices that promote themselves in the eyes of others or promote a particular group or viewpoint etc. Such deliberate partisanship may be recognised as such, sometimes easily, sometimes with difficulty. Much more difficult to recognise are the skills, the values, the beliefs, the interests, the prejudices and the biases that are at the heart of a particular society and culture, that are fundamental to the way it operates. Indeed, they are assumed and supported by that society and culture. Theories emerge that support particular world-views.

The key challenge at the heart of Functional Collaboration is the 'subtle and difficult' (Bohm 2002: 22) terrain of the subject as subject, a terrain which requires the long slow process of self-awareness, self-understanding, self-affirmation and self-appropriation of the operative dynamics of the subject as subject: the desire to understand, the desire to create something worthwhile, the desire for sociality (for relationships, for co-operation and for solidarity) and the desire to love and to be loved. It is the dynamic of the desire to understand which grounds the set of eight questions that constitute Functional Collaboration as a scientific approach to housing, their functional relationships, their anticipated answers and their respective methods. It is the dynamic of the desire to create something worthwhile that makes these questions cyclic. It is the dynamic of the desire for sociality that shifts these cyclic inter-related questions from individual operations to a framework for collaborative creativity among all researchers.

The future of housing research depends upon housing researchers meeting this key challenge as it now emerges in history. If we are to go beyond partisan understandings of housing and partisan policies, housing researchers have the personal challenge of distinguishing between different types of questions, their anticipated answers, their ongoing demands for adequate answers and the ongoing development of methods by which each type of question is answered.

Functional Collaboration seeks to meet the set of problems that confront housing research by dividing up the work according to differing questions and methods for answering them. It is these different types of questions that ground the distinctions between the functional specialties, their relationship to one another and the operations that constitute each of the functional specialties. Without understanding this, Functional Collaboration is but a series of arbitrary though useful stages in a process.

At the outset I proposed that Functional Collaboration was a scientific approach to housing, that it was constituted by a complete set of eight inter-related questions. To envisage it as such a set required a re-orientation of the methods that characterise housing research. This re-orientation sought to select those elements of each method that were essential, significant and relevant to the method and thus constituted the method. At the same time, this re-orientation set aside associated unessential, insignificant and irrelevant elements. In this way, I came to characterise each of the methods as a paradigm shift. It is not simply a shift in each method but rather a complex of shifts, each in relation to one another.

Research proposes a shift in current housing research *from* answering any type of question raised, *to* answering an empirical question: what events are occurring in this time and place and to what extent are these events associated? It answers this question with a view to providing data or evidence for Interpretation to answer the theoretical question: what are the elements and their relationships that constitute housing? The empirical question anticipates the occurrence of events. It anticipates a frequency, participant descriptions of feelings, beliefs, attitudes etc., descriptions of sequences of events and the association of the occurrence of events or properties or characteristics.

Interpretation proposes a shift in the way housing is defined *from* descriptive definitions *to* answering a theoretical question: what are the elements and their relationships that constitute housing? The answer is an explanatory definition, an ordered set of sets of related variable elements that constitutes housing. This explanatory definition distinguishes between those elements which are essential, significant and relevant to the constitution of housing and those which are unessential, insignificant and irrelevant to the constitution of housing. Housing is used for differing purposes by different groups. It is these roles or functions associated with housing which order the particularity of the set of related variable elements that constitute housing and create an actual operating housing system.

History proposes a shift in history *from* documenting and recounting a series of events in some remote or near past *to* answering a historical question: what dynamic or vector has provoked ongoing change in an actual operating housing system? Within History, a theory of housing is a heuristic for grasping an actual operating housing system as a complete ordered set of sets of related variable elements that constitute housing. The role of this heuristic is to identify sequences of actual operating housing systems. The function of History is to identify the complex operative dynamic or vector that changes one actual operating housing system into another.

Dialectic proposes a shift in evaluation *from* simply pointing to the limitations of some state of affairs (whether that be an existing policy, programme or project, or the exercise of dominating power by individuals and groups acting in their own interest as they aggrandise themselves by alienating others) *to* answering an evaluative/critical question: what is the best of the past? Dialectic evaluates the vectors of the history of housing by understanding how this history came about, appreciating what was achieved and critiquing its limitations and integrating or relating the best of the past within the horizon of the subject as subject (and the dynamics of the subject).

Foundations proposes a shift in the foundations of research *from* a horizon based on arbitrary axioms, self-evident truths or first principles *to* answering the transformative/visionary question: who will I be (as a housing researcher) or, who will we be (as housing researchers)? Foundations locates the foundations of research in the transformative decision of the researcher to operate in accord with the dynamics of the subject as subject, in particular, in the dynamics of the desire to understand. Foundations thematises the subject as subject and this transformative decision.

Policies proposes a shift *from* a focus on government or organisational policy – policies whose range is limited to one or other purpose which serve the interests of that government or organisation – *to* answering the policy question: what new operative dynamic or vector will best promote the future development of the current actual operating housing system into future systems? The answer to this question is a new direction which incorporates the totality of housing purposes rather than one or other purpose which continues the self-aggrandisement of individuals and groups.

Systematics proposes a shift *from* a focus on how a government or an organisation will use housing to achieve its particular purposes in the context of other competing purposes *to* answering the strategic question: what course of action will integrate a new vector or direction within the complex series of contexts that constitute an actual operative housing system? The answer to this question will integrate a new housing policy direction within its technological, economic, social, political and cultural contexts.

Communications proposes a shift *from* implementation as a top-down command process *to* asking the practical question: what practices/activities in this time and place will achieve a strategic course of action which will realise a new actual operative housing system? The answer to this question will identify particular activities/practices which will be institutionalised into the roles and practices of individuals and groups.

Functional Collaboration gives a more precise meaning to the range of questions to which we need new answers before we can project our way forward. Indeed, it is a complete set of inter-related questions that constitute 'cumulative and progressive results'. It envisages an integration of the methods that constitute science: empirical, theoretical, historical, evaluative/critical, transformative/ visionary, policy, strategic and practical.

Functional Collaboration will slowly and globally address the problem of fragmentation and complexity. As global in scope, each functional specialty can be undertaken by anyone anywhere; researchers can take the results of a prior functional specialty as a starting point, wherever it occurred, and pass on the results of their work to another functional specialty. Functional Collaboration draws on the resources and experiences of all societies and cultures to find solutions to local problems. It reaches back into history to grasp the beginnings of new vectors of change. It respects the achievements of all societies and cultures (while it may also recognise their (even severe) limitations) and draws on this experience. Through the implementation functional specialties, it reaches forward to transformations within each particular society and culture (in all their differences) through the enlightened co-operation of its people.

Functional Collaboration is a transdisciplinary, transcultural solution to the emerging problems in science. It addresses the complex issues raised by Lawrence and Després. To repeat parts of the quote from them at the beginning of the last chapter: it accepts 'local contexts and uncertainty'; it deals 'with real-world topics' and generates 'knowledge that not only address societal problems but also contribute to their solution'; and it bridges 'the gap between knowledge derived from research and decision-making processes in society'.

It takes some work to grasp the role of each functional specialty and Functional Collaboration as a whole. It will only come in the process of doing empirical work. Flyvbjerg (2001: 145) says of phronesis that '[t]he promise of this methodology is better understood through examining cases of its employment in their entirety, or, even better, by employing the methodology oneself in an actual study'. Something similar may be said of Functional Collaboration. Until attempts are made to implement it, its promises remain unknown and its potential remains unfulfilled.

We live in a society where the study of housing is dominated by everyday common-sense thinking with its demand for immediate practical solutions (particularly those that can be initiated by governments) and where the division of labour and specialisation based on disciplines and on data areas is thoroughly embedded in the culture of academia. Housing research is beginning to recognise some of the problems with current research. Within the current environment, however, the implementation of Functional Collaboration will not be easy. An attempt at the integration of different genres of housing research mark a new beginning but the implementation of Functional Collaboration is still a long way off. Before it can be implemented, Functional Collaboration has to be considered worthwhile (something better than what we have now). Before it is considered worthwhile, it has to be understood. Before it can be understood, researchers have to recognise the problems with current housing research.

It will require 'fantasy and lateral thinking' in which we have to imagine the academy of the future, a successful collaborative culture, the restructuring of research organisations such as universities. Functional Collaboration proposes a new division of labour which involves a different way of thinking, a different way of relating to other researchers and decision-makers, a different way of organising the products of research (libraries, databases and journals), a different way of organising the academy, conferences, institutes etc., different styles of communication etc. Rather than being structured around disciplines and data areas, housing researchers would be structured around the functional specialties, 'all structured by the common achieved or to-be-achieved meaning of its members' (Braio 2000: 57). The internal dialogues between the functional specialties will be

> remote from common-sense discourse. Yet it will be publicly engaging and effective in its images, its symbols and the affective, intersubjective responses they call forth. On the other hand, its initial remoteness from public discourse is, of course, one of the difficult and perhaps unacceptable challenges of the future ... Still, the functionally specialized withdrawal from undifferentiated consciousness is not extrinsic to such a community's effort. Rather, it is

required if the return through the eighth functional specialty, communications, seriously is to shift and transform the cultures of the globe.

(Braio 2000: 57–58)

Beyond housing research, Functional Collaboration does indeed have a far larger role in 'transforming the cultures of the globe'. It points to how we make progress in any field of human endeavour by asking and finding new answers to eight inter-related questions. Its larger context is the challenges that humanity faces; our current inability to deal with the environmental, financial, economic, social, political and cultural crises of our time. No longer can we isolate ourselves from the issues that confront the future of earth. As Stewart Brand in his *Whole Earth Discipline: An Ecopragmatist Manifesto* (2010: 272) notes:

Whether it's called managing the commons, natural-infrastructure maintenance, tending the wild, niche construction, ecosystem engineering, mega-gardening, or intentional Gaia, humanity is now stuck with a planet stewardship role...If humanity's role has expanded to the point that the entire Earth is our niche, the trend of the changes we have made lately indicates we are doing a poor job of niche maintenance.

We have yet to develop an effective framework for collectively taking responsibility for this whole earth stewardship role. Functional Collaboration points to new possibilities.

Here I am making a beginning. I am simply asking current housing researchers to differentiate between the different types of questions they ask. The implementation of Functional Collaboration is a vast enterprise. I am looking forward to a time when the reciprocal dependence of the functional specialties that constitute Functional Collaboration are understood and appreciated. As the epigraph inside the front cover notes: 'Not only is that understanding required; one has to be familiar with what is called the *acquis*, what has been settled, what no one has any doubt about at the present time' (Lonergan [1968] 2010: 462).

The challenge now is for housing researchers to grasp the dynamics of the subject as subject as the grounds for Functional Collaboration and a scientific approach to housing, a scientific approach which could provide housing researchers with a framework for collaborative creativity that will relate and integrate disparate types of current research and will present decision-makers with practical advice on future directions for housing. Then, we can confront the problems that beset housing and look forward not just to random *ad hoc* development but rather to making consistent, ongoing and cumulative progress in housing.

Bibliography

ABS (2009) *Information Paper: Survey of Income and Housing, User Guide, Australia, 2007–08*, Australian Bureau of Statistics, Canberrra, viewed 29 June 2010, http://www.abs.gov.au/AUSSTATS/subscriber.nsf/log?openagent&65530_2007-08.pdf&6553.0&Publication&88CD2DFD5A9D84B5CA25761700193589&&2007-08&20.08.2009&Latest.

——(2011a) *2011 Census of Population and Housing: Community Profile Series, Basic Community Profile, Australia*, Australian Bureau of Statistics, Canberra, viewed 10 November 2012, http://www.abs.gov.au/websitedbs/censushome.nsf/home/community profiles?opendocument&navpos=230.

——(2011b) *Census Dictionary: Australia, 2011,* Australian Bureau of Statistics, Canberra, viewed 12 April 2013 http://www.abs.gov.au/ausstats/abs@.nsf/Lookup/2901.0Main%20Features802011/$FILE/2011%20Census%20Dictionary%2027102011.pdf.

ACOSS (1992) *National Project on the Management of Social Housing: Final Report*, Australian Council of Social Service, Sydney.

——(2008) *Submission to the Senate Inquiry into Housing Affordability*, Australian Council of Social Service, Sydney, viewed 10 February 2010, http://acoss.org.au/images/uploads/ACOSS_Senate_Inquiry_submission_final.pdf.

Adams J (2012) 'Collaborations: The Rise of Research Networks', *Nature*, 490, 7420, 335–336.

——(2013) 'Collaborations: The Fourth Age of Research', *Nature*, 497, 7451557–560.

AIHW (2009) *Australia's Welfare 2009*, Australian Institute of Health and Welfare, Canberra, viewed 10 February 2010, http://www.aihw.gov.au/publications/aus/aw09/aw09.pdf.

——(2010) *Public Rental Housing 2008–09*, Australian Institute of Health and Welfare, Canberra, viewed 10 February 2010, http://www.aihw.gov.au/publications/hou/218/11420.pdf.

Ali S (1985) *Preliminary Discussion Report on a Study of Public Rental Housing Subsidies: Market Rent, Cost Rent and Rebated Rent: A Question of Pricing Policies*, Review Branch, Research and Review Unit, Ministry of Housing, Melbourne.

Allen C (2005) 'On the Social Relations of Contract Research Production: Power, Positionality and Epistemology in Housing and Urban Research', *Housing Studies*, 20, 6, 989–1007.

——(2009) 'The Fallacy of "Housing Studies": Philosophical Problems of Knowledge and Understanding in Housing Research', *Housing, Theory and Society*, 26, 1, 53–79.

Ambrose P (1991) 'The Housing Provision Chain as a Comparative Analytical Framework', *Housing, Theory and Society*, 8, 2, 91–104.

——(1992) 'The Performance of National Housing Systems a Three-Nation Comparison', *Housing Studies*, 7, 3, 163–176.

Anderson B & McShane P (2002) *Beyond Establishment Economics: No Thank-You Mankiw*, Axial, Halifax.

Aristotle ([c.322 BC] 2007a) *Posterior Analytics (or Analytica Posteriora)*, trans. GRG Mure, eBooks@Adelaide, viewed 11 January 2011, http://ebooks.adelaide.edu.au/a/aristotle/a8poa/a8poa.epub.

——([c.322 BC] 2007b) *Physics*, trans. RP Hardie and RK Gaye, eBooks@Adelaide, viewed 9 June 2011, http://ebooks.adelaide.edu.au/a/aristotle/physics/.

Arthurson K & Jacobs K (2004) 'A Critique of the Concept of Social Exclusion and Its Utility for Australian Social Housing Policy', *Australian Journal of Social Issues*, 39, 1, 25–40.

Atkinson R (2008) *Housing Policies, Social Mix and Community Outcomes*, Final Report 122, AHURI, Melbourne, viewed 4 February 2010, http://www.ahuri.edu.au/publications/download/40500_fr.

Atkinson R & Jacobs K (2008) *Public Housing in Australia: Stigma, Home and Opportunity*, Discussion Paper No 1, Housing and Community Research Unit, School of Sociology, University of Tasmania, Hobart, viewed 4 February 2010, http://www.utas.edu.au/sociology/HACRU/HACRU_Discussion_Paper_1_Public_housing_in_Australia_-_Stigma,_home_and_opportunity.pdf.

——(2009) 'The Social Forces and Politics of Housing Research: Reflections from within the Academy', *Housing, Theory and Society*, 26, 4, 233–247.

Australian Institute of Urban Studies (1975) *Housing for Australia: Philosophy and Policies: The Report of a Task Force of the Australian Institute of Urban Studies*, Publication 58, Australian Institute of Urban Studies, Canberra.

Bachelard G (1969) *The Poetics of Space*, Beacon, Boston.

Bacon F ([1620] 2005) *The New Organon or True Directions Concerning the Interpretation of Nature*, Constitution Society, Austin, Texas, viewed 10 March 2011, http://www.constitution.org/bacon/nov_org.htm.

Ball M (1983) *Housing Policy and Economic Power: The Political Economy of Owner Occupation*, Methuen, London.

——(1986) 'Housing Analysis: Time for a Theoretical Refocus', *Housing Studies*, 1, 3, 147–165.

——(1998) 'Institutions in British Property Research: A Review', *Urban Studies*, 35, 9, 1501–1517.

Ball M & Harloe M (1992) 'Rhetorical Barriers to Understanding Housing Provision: What the "Provision Thesis" Is and Is Not', *Housing Studies*, 7, 1, 3–15.

Ball M, Harloe M & Martens M (1988) *Housing and Social Change in Europe and the USA*, Routledge, London.

Barnett FO (1933) *The Unsuspected Slums: An Illustrated Summary of a Thesis Submitted to the Melbourne University Surveying the Slum Problem of Melbourne*, Herald Press, Melbourne.

Barnett FO & Burt WO (1942) *Housing the Australian Nation*, Left Book Club of Victoria, Melbourne.

Barnett FO, Burt WO & Heath F (1944) *We Must Go On: A Study in Planned Reconstruction and Housing*, Book Depot, Melbourne.

Batten DC (1999) 'The Mismatch Argument: The Construction of a Housing Orthodoxy in Australia', *Urban Studies*, 36, 1, 137–151.

Baumgardner J & Richards A (2003) 'The Number One Question About Feminism', *Feminist Studies*, 29, 2, 448–452.

Beer P (2009) *An Introduction to Bernard Lonergan: Exploring Lonergan's approach to the great philosophical questions*, Sid Harta Publishers, Glen Waverley.

Bengtsson B (1995) 'Politics and Housing Markets: Four Normative Arguments', *Housing, Theory and Society*, 12, 3, 123–140.

——(1998) 'Tenants' Dilemma: On Collective Action in Housing', *Housing Studies*, 13, 1, 99–120.

——(2009a) 'Political Science as the Missing Link in Housing Studies', *Housing, Theory and Society*, 26, 1, 10–25.

——(2009b) *Political Science, Politics, and Housing Studies ISA International Housing Conference, Housing Assets Housing People*, University of Glasgow, viewed 25 October 2012 http://www.gla.ac.uk/media/media_129760_en.pdf.

Benton J (2008) *Shaping the Future of Language Studies*, Axial, South Brookfield.

Benton J, Drage A & McShane P (2005) *Introducing Critical Thinking*, Axial, Cape Breton.

Berger PL & Luckmann T (1971) *The Social Construction of Reality: A Treatise in the Sociology of Knowledge*, Penguin, Harmondsworth.

Berry M (1979) *Marxist Approaches to the Housing Question*, Centre for Urban and Regional Studies, University of Birmingham, Birmingham.

——(1983) 'Posing the Housing Question in Australia: Elements of a Theoretical Framework for a Marxist Analysis of Housing', in L Sandercock and M Berry (eds) *Urban Political Economy: The Australian Case*, George Allen and Unwin, Sydney.

——(1986) 'Housing Provision and Class Relations under Capitalism: Some Implications of Recent Marxist Class Analysis', *Housing Studies*, 1, 2, 109–121.

——(1988) 'To Buy or Rent? The Demise of a Dual Tenure Housing Policy in Australia, 1945–60', in R Howe (ed.) *New Houses for Old: Fifty Years of Public Housing in Victoria 1938–1988*, Ministry of Housing and Construction, Melbourne, pp. 95–122.

Berry M & Hall J (2002) *New Approaches to Expanding the Supply of Affordable Housing in Australia: An Increasing Role for the Private Sector*, Final Report 14, AHURI, Melbourne, viewed 1 March 2011, http://www.ahuri.edu.au/publications/download/30021_fr.

Berry M, Hall J & Allen Consulting Group (2001) *Policy Options for Stimulating Private Sector Investment in Affordable Housing across Australia: Stage 2 Report: Identifying and Evaluating the Policy Options*, AHURI, Melbourne, viewed 1 March 2011, http://www.ahuri.edu.au/publications/download/30063_stage2.

Berry M, Whitehead C, Williams P & Yates J (2004) *Financing Affordable Housing: A Critical Comparative Review of the United Kingdom and Australia*, Final Report, AHURI, Melbourne, viewed 28 September 2010, http://www.ahuri.edu.au/publications/download/30206_fr.

Bhaskar R (2008) *A Realist Theory of Science*, Taylor & Francis, Hoboken.

Bisset H & Victorian Council of Churches (1991) *Housing Affordability and Housing Finance: A Submission to the National Housing Strategy in Response to Issues Papers 2 & 3*, Ecumenical Housing Unit, Melbourne.

Black J, Hashimzade N & Myles G (2009) 'IS–LM Model', *Dictionary of Economics*, Oxford Reference Online, Oxford University Press, viewed 7 June 2011, http://www.oxfordreference.com/views/ENTRY.html?subview=Main&entry=t19.e1707.

Blaikie NWH (2003) *Analyzing Quantitative Data: From Description to Explanation*, Sage, London.

——(2007) *Approaches to Social Enquiry: Advancing Knowledge*, Polity, Cambridge.

Blaug M (1995) 'Why Is the Quantity Theory of Money the Oldest Surviving Theory in Economics?', in M Blaug, D Pantikin, G Wood, D O'Brien and WA Eltis (eds) *The Quantity Theory of Money: From Locke to Keynes and Friedman*, Edward Elgar, Aldershot, pp. 27–49.

Blunt A & Dowling R (2006) *Home,* Taylor & Francis, Hoboken.

Bohm D (2002) *Wholeness and the Implicate Order*, Routledge, Hoboken.

Bourdieu, P (1979) 'The Kayble House or the World Reversed' in *Algeria 1960*, Cambridge University Press, Cambridge, 133–153.

Boyne GA & Walker RM (1999) 'Social Housing Reforms in England and Wales: A Public Choice Evaluation', *Urban Studies (Routledge)*, 36, 13, 2237–2262.

Bradley Q (2008) 'Capturing the Castle: Tenant Governance in Social Housing Companies', *Housing Studies*, 23, 6, 879–897.

Brand S (2010) *Whole Earth Discipline: An Ecopragmatist Manifesto*, ebook edition, Atlantic Books, London.

Braio FP (2000) 'The 'Far Larger' Work of Insight's Epilogue', *Lonergan Workshop: Lonergan and the Human Sciences*, 16, 49–66, viewed 8 February 2011, http://www.lonerganresource.com/pdf/journals/Lonergan_Workshop_Vol_16.pdf.

Bramley G (1991) *Public Sector Housing Rents and Subsidies: Alternative Approaches and Their Applications to Selected Localities*, Working Paper 92, School of Advanced Urban Studies, University of Bristol, Bristol.

Bridge C, Davy L, Judd B, Flatau P, Morris A & Phibbs P (2011) *Age-Specific Housing and Care for Low to Moderate Income Older People*, Final Report 174, AHURI, Melbourne, viewed 24 October 2011, http://www.ahuri.edu.au/publications/download/70589_fr.

Browne G & Courtney M (2005) 'Housing, Social Support and People with Schizophrenia: A Grounded Theory Study', *Issues in Mental Health Nursing*, 26, 3, 311–326.

Building South Australia (2005) *Housing Plan for South Australia*, Adelaide, viewed 3 March 2011, http://blogs.dfc.sa.gov.au/m/dfcweb_hsa/501/download.aspx.

Bunnin N & Yu J (eds) (2007) *The Blackwell Dictionary of Western Philosophy*, John Wiley, Chichester.

Burke T (1993) *International Low Income Housing Systems: A Report Prepared for the Industry Commission's Public Housing Inquiry*, Centre for Urban and Social Research, Swinburne University of Technology, Melbourne.

Burke T & Hulse K (2003) *Allocating Social Housing*, Positioning Paper 47, AHURI, Melbourne, viewed 3 February 2010, http://www.ahuri.edu.au/publications/download/50141_pp.

——(2010) 'The Institutional Structure of Housing and the Sub-Prime Crisis: An Australian Case Study', *Housing Studies*, 25, 6, 821–838.

Burke T & Zakharov R (2005) *Long-Term Housing Futures for Australia: Using 'Foresight' to Explore Alternative Visions and Choices*, Final Report 77, AHURI, Melbourne, viewed 9 February 2010, http://www.ahuri.edu.au/publications/download/50225_fr.

Burke T, Pinnegar S, Phibbs P, Neske C, Gabriel M, Ralston L & Ruming K (2007) *Experiencing the Housing Affordability Problem: Blocked Aspirations, Trade-Offs and Financial Hardships*, National Research Venture 3: Housing Affordability for Lower Income Australians: Research Paper 9, AHURI, Melbourne, viewed 12 June 2010, http://www.ahuri.edu.au/publications/download/nrv3_research_paper_9.

Burke T, Slaughter R & Voros J (2004) *Long-Term Housing Futures for Australia: Using 'Foresight' to Explore Alternative Visions and Choices*, Positioning Paper 73, AHURI, Melbourne, viewed 22 February 2011, http://www.ahuri.edu.au/publications/download/50225_pp.

Bury JB (1955) *The Idea of Progress*, Dover, New York.

Butterfield H (1957) *The Origins of Modern Science 1300–1800*, Macmillan, New York.

Byrne PH (2002) 'Research: An Illustration from Galileo Studies', *Method: A Journal of Lonergan Studies*, 20, 1, 21–32, viewed 18 June 2010, http://www.lonerganresource.com/pdf/journals/Method_Vol_20_No_1.pdf.

Carter RA (1980a) 'Housing Policy in the 1970s', in RB Scotton and H Ferber (eds) *Public Expenditure and Social Policy in Australia, Volume II: The First Fraser Years, 1976–1978*, Longman Cheshire, Melbourne, pp. 77–143.

——(1980b) 'The Effect of Commonwealth Housing Policies on Lower Income Groups, 1972 to 1980', *Australian Economic Review*, 13, 3, 18–23.

City of Boroondara (2007) *Social Housing Policy 2008–2011*, City of Boroondara, Melbourne, viewed 3 March 2011, http://boroondara.vic.gov.au/your_council/local-laws-policies/community/social-housing.

City of Knox (2010) *Knox Affordable Housing Action Plan 2007–2012 (Updated October 2010)*, City of Knox, Melbourne, viewed 3 March 2011, http://www.knox.vic.gov.au/Files/Knox_Affordable_Housing_Action_Plan_2007_-_2012_updated_2010.pdf.

City of Melbourne (*c.*2001) *Linking People, Homes and Communities: A Social Housing Strategy 2001–2004*, City of Melbourne, Melbourne, viewed 3 March 2011, http://unpan1.un.org/intradoc/groups/public/documents/apcity/unpan011804.pdf.

——(2006) *Social and Affordable Housing: Framework 2006–2009: Housing for Everyone*, City of Melbourne, Melbourne, viewed 3 March 2011, http://www.melbourne.vic.gov.au/CommunityServices/SocialSupport/Documents/framework_socialaffordable_housing_2006_2009.pdf.

City of Sydney (2009) *Draft Affordable Rental Housing Strategy 2009–2014*, City of Sydney, Sydney, viewed 3 March 2011, http://www.cityofsydney.nsw.gov.au/ Council/documents/OnExhibition/DraftCityAffordableHousingStrategy2009-14/AttachmentA-AffordableRentalHousingStrategyFeb2009.pdf.

Clapham D (1997) 'The Social Construction of Housing Management Research', *Urban Studies*, 34, 5/6, 761–774.

——(2002) 'Housing Pathways: A Post Modern Analytical Framework', *Housing, Theory and Society*, 19, 2, 57–68.

——(2004) 'Housing Pathways: A Social Constructionist Research Framework', in K Jacobs, J Kemeny and T Manzi (eds) *Social Constructionism in Housing Research*, Ashgate, Aldershot, pp. 93–116.

——(2005) *The Meaning of Housing: A Pathways Approach*, Policy Press, Bristol.

——(2009) 'Introduction to the Special Issue: A Theory of Housing: Problems and Potential', *Housing, Theory and Society*, 26, 1, 1–9.

Cloud W & Roll S (2011) 'Denver Housing Authority's Park Avenue HOPE VI revitalization project: community impact results', *Housing Policy Debate*, 21, 2, 191–214.

Cole I (2006) 'Hidden from History? Housing Studies, the Perpetual Present and the Case of Social Housing in Britain', *Housing Studies*, 21, 2, 283–295.

Commission of Inquiry into Poverty (1975) *Poverty in Australia: First Main Report Volume I*, Australian Government Publishing Service, Canberra.

Commonwealth Housing Commission (1944) *Final Report*, Ministry of Post-War Reconstruction, Canberra.

Corrigan P (1997) *The Sociology of Consumption: An Introduction*, Sage, London.

Crotty M (1998) *The Foundations of Social Research: Meaning and Perspective in the Research Process*, Allen & Unwin, Sydney.

Crowe FE (1973) 'Early Jottings on Bernard Lonergan's *Method in Theology*', *Science et Esprit*, 25, 121–138.

Dalton T (2009) 'Housing Policy Retrenchment: Australia and Canada Compared', *Urban Studies*, 46, 1, 63–91.

Danermark B, Ekström M, Jakobsen L & Karlsson JC (2002) *Explaining Society: An Introduction to Critical Realism in the Social Sciences*, Routledge, London.

Dargavel R & Kendig H (1986) 'Political Rhetoric and Program Drift: House and Senate Debates on the Aged or Disabled Persons' Homes Act', *Australian Journal on Ageing*, 5, 2, 23–31.

Daunton MJ (1984) *Councillors and Tenants: Local Authority Housing in English Cities, 1919–1939*, Leicester University Press, Leicester.

——(1987) *A Property-owning Democracy?: Housing in Britain*, Faber, London.

——(1990) *Housing the Workers, 1850–1914: A Comparative Perspective*, Leicester University Press, London.

——(1991) 'Health and Housing in Victorian London', *Medical History*, Supplement 11, 126–144.

Davies HTO, Nutley SM & Smith PC (eds) (2000) *What Works? Evidence-Based Policy and Practice in Public Services*, Policy Press, Bristol.

Davison G & Fincher R (1998) 'Urban Studies in Australia: A Road Map and Ways Ahead', *Urban Policy and Research*, 16, 3, 183–197.

De Vaus DA (2002) *Surveys in Social Research*, Allen & Unwin, Sydney.

De Vroey M (2000) 'IS–LM 'à la Hicks' Versus IS–LM 'à la Modigliani'', *History of Political Economy*, 32, 2, 293–316.

Dewey J (1938) *Logic: The Theory of Inquiry*, Holt, Rinehart and Winston, New York.

Dockery AM, Feeny S, Hulse K, Ong R, Saugeres L, Spong H, Whelan S & Wood G (2008) *Housing Assistance and Economic Participation*, National Research Venture 1: Housing Assistance and Economic Participation, Final Research Paper, AHURI, Melbourne, viewed 10 June 2010, http://www.ahuri.edu.au/publications/download/nrv1_final_research_paper.

Dodson J (2006) 'The 'Roll' of the State: Government, Neoliberalism and Housing Assistance in Four Advanced Economies', *Housing, Theory and Society*, 23, 4, 224–243.

——(2007) *Government Discourse and Housing*, Ashgate, Aldershot.

Dol K & Haffner M (eds) (2010) *Housing Statistics in the European Union 2010*, Ministry of the Interior and Kingdom Relations, The Hague, viewed 13 October 2012 http://abonneren.rijksoverheid.nl/media/dirs/436/data/housing_statistics_in_the_european_union_2010.pdf.

Dovey K (1992) 'Model Houses and Housing Ideology in Australia', *Housing Studies*, 7, 3, 177–188.

Dowling R & Mee K (2007) 'Home and Homemaking in Contemporary Australia', *Housing, Theory and Society*, 24, 3, 161–165.

Drage A (2003) 'Philip McShane's Axial Period: An Interpretation', *Journal of Macrodynamic Analysis*, 4, 128–179, viewed 8 June 2013, http://journals.library.mun.ca/ojs/index.php/jmda/article/view/142/95.

——(2005) *Thinking Woman*, Axial, Cape Breton.

Dunleavy P (1981) *The Politics of Mass Housing in Britain, 1945–1975: A Study of Corporate Power and Professional Influence in the Welfare State*, Clarendon Press, Oxford.

Easthope H (2004) 'A Place Called Home', *Housing, Theory and Society*, 21, 3, 128–138.

Eather W (1988) 'We Only Build Houses: The Commission 1945–60', in R Howe (ed.) *New Houses for Old: Fifty Years of Public Housing in Victoria 1938–1988*, Ministry of Housing and Construction, Melbourne, pp. 69–94.

Econsult (Australia) (1989) *Housing Allowances in the Australian Context: Market Impacts and Cost Effectiveness*, National Housing Policy Review Background Paper 1, Department of Community Services and Health, Canberra.

Ecumenical Housing (1997) *National Housing Policy: Reform and Social Justice: An Alternative Approach to the Housing Assistance Reform Debate*, Ecumenical Housing, Melbourne.

Ecumenical Housing, Property Concept & Management Pty Ltd & Urban Land Corporation (1999) *ACT Housing Multi-Unit Property Plan*, Ecumenical Housing, Melbourne.

Esping-Andersen G 1990, *The Three Worlds of Welfare Capitalism*, Polity, Cambridge.

Fairclough N (1989) *Language and Power*, Longman, London.

——(1993) *Discourse and Social Change*, Polity, Cambridge.

——(1995) *Critical Discourse Analysis: The Critical Study of Language*, Longman, London.

Feyerabend PK (1993) *Against Method: Outline of an Anarchistic Theory of Knowledge*, Verso, London.

Feynman R (1969) 'What Is Science?', *Physics Teacher*, 7, 6, 313–320.

Fischer F, Miller G & Sidney M (2007) *Handbook of Public Policy Analysis: Theory, Politics, and Methods*, CRC Press, Boca Raton.

Flanagan J (1997) *Quest for Self-Knowledge: An Essay in Lonergan's Philosophy*, University of Toronto Press, Toronto.

Flick U, Steinke I & von Kardoff E (2004) 'What Is Qualitative Research? An Introduction to the Field', in U Flick, I Steinke and E von Kardoff (eds) *A Companion to Qualitative Research*, Sage, London, pp. 3–11.

Flood J (1989) *Financing Public Housing: The Needs, the Options and the Risks*, National Housing Policy Review Background Paper 4, Department of Community Services and Health, Canberra.

Flyvbjerg B (2001) *Making Social Science Matter: Why Social Inquiry Fails and How It Can Succeed Again*, Cambridge University Press, Cambridge.

Forster EM ([1927] 1974) *Aspects of the Novel and Related Writings*, Edward Arnold, London.

Friend JK, Power JM & Yewlett CJL (1974) *Public Planning: The Inter-Corporate Dimension*, Tavistock, London.

Gabriel M & Jacobs K (2008) 'The Post-Social Turn: Challenges for Housing Research', *Housing Studies*, 23, 4, 527–540.

Gadamer H-G (1975) *Truth and Method*, Continuum, New York.

Garfinkel H (1962) 'Common Sense Knowledge of Social Structures: The Documentary Method of Interpretation', in JM Scher (ed.) *Theories of the Mind*, Free Press of Glencoe, New York, pp. 689–712.

——(1967) *Studies in Ethnomethodology*, Prentice-Hall, Englewood Cliffs.

Gervais D (2001) 'Englands of the Mind', *Cambridge Quarterly*, 30, 2, 151–168.

Gibb K (2002) 'Trends and Change in Social Housing Finance and Provision Within the European Union', *Housing Studies*, 17, 2, 325–336.

——(2009) 'Housing Studies and the Role of Economic Theory: An (Applied) Disciplinary Perspective', *Housing, Theory and Society*, 26, 1, 26–40.

Giddens A (1979) *Central Problems in Social Theory: Action, Structure and Contradiction in Social Analysis*, Macmillan, London.

——(1993) *New Rules of Sociological Method: A Positive Critique of Interpretative Sociologies*, Polity Press, Cambridge.

Gillis-Drage A (2010) 'The Number One Question About Feminism: The Third Wave and the Next Half-Century', *Journal of Macrodynamic Analysis*, 5, 7–19, viewed 6 June 2013, http://journals.library.mun.ca/ojs/index.php/jmda/article/view/178/123.

Goffman E (1961) *Encounters: Two Studies in the Sociology of Interaction*, Bobbs-Merrill, Indianapolis.

——(1969) *The Presentation of Self in Everyday Life*, Allen Lane, London.

Greenberg J 1978, *Universals of Human Language*, Stanford University Press, Stanford.

Grey A, Hepworth NP & Odling-Smee J (1981) *Housing Rents, Costs and Subsidies: A Discussion Document*, Chartered Institute of Public Finance and Accountancy, London.

Grosz E (2000) 'Histories of a Feminist Future', *Signs: Journal of Women in Culture and Society*, 25, 4, 1017–1021.

Gruis V & Nieboer N (2004) 'Strategic Housing Management: An Asset Management Model for Social Landlords', *Property Management,*, 22, 3, 201–213.

Gurran N, Milligan V, Baker D & Bugg LB (2007) *International Practice in Planning for Affordable Housing: Lessons for Australia*, Positioning Paper 99, AHURI, Melbourne, viewed 26 September 2010, http://www.ahuri.edu.au/publications/download/60322_pp.

——(2008) *New Directions in Planning for Affordable Housing: Australian and International Evidence and Implications*, Final Report 120, AHURI, Melbourne, viewed 26 September 2010, http://www.ahuri.edu.au/publications/download/60322_fr.

Hall J & Berry M (2004) *Operating Deficits and Public Housing: Policy Options for Reversing the Trend*, Final Report 55, AHURI, Melbourne, viewed 4 February 2010, http://www.ahuri.edu.au/publications/download/30154_fr.

——(2009) *Operating Deficits and Community Housing: Policy Options for Reversing the Trend*, Final Report 126, AHURI, Melbourne, viewed 4 February 2010, http://www.ahuri.edu.au/publications/download/30355_fr.

Harloe M (1995) *The People's Home? Social Rented Housing in Europe & America*, Blackwell, Oxford.

Harris D (1988) 'Not Above Politics: Housing Reform in Melbourne 1910–29', in R Howe (ed.) *New Houses for Old: Fifty Years of Public Housing in Victoria 1938–1988*, Ministry of Housing and Construction, Melbourne, pp. 1–19.

Hayward D (1986) 'The Great Australian Dream Reconsidered: A Review of Kemeny', *Housing Studies*, 1, 4, 210–219.

——(1996) 'The Reluctant Landlords? The History of Public Housing in Australia', *Urban Policy and Research*, 14, 1, 5–35.

Hefling CC (2000) *Why Doctrines?*, Lonergan Institute at Boston College, Chestnut Hill, Mass.

Hegel GWF ([1805–6] 1955) *Lectures on the History of Philosophy*, trans. ES Haldane, viewed 11 November 2010, http://www.marxists.org/reference/archive/hegel/works/hp/hpconten.htm.

Heidegger M ([1951] 1975) *Poetry, Language, Thought,* HarperCollins, New York.

——(1973) *Being and Time*, Basil Blackwell, Oxford.

——(1988) *The Basic Problems of Phenomenology*, Indiana University Press, Bloomington.

Heilbroner RL (1973) 'Economics as a 'Value-Free' Science', *Social Research*, 40, 1, 129–143.

Hill MJ (1997) *The Policy Process in the Modern State*, Prentice Hall, New York.

——(2003) *Understanding Social Policy*, Blackwell, Malden.

Hill MJ & Hupe PL (2002) *Implementing Public Policy: Governance in Theory and in Practice*, Sage, London.

Hill O (1883) *Homes of the London Poor*, London, viewed 24 October 2011, http://www.victorianlondon.org/publications/homesofthelondonpoor.htm.

Hills J (1988) *Twenty-First Century Housing Subsidies: Durable Rent-Fixing and Subsidy Arrangements for Social Housing*, London School of Economics, London.

——(1991) *Unravelling Housing Finance: Subsidies, Benefits and Taxation* Clarendon Press, Oxford.

——(2007) *Ends and Means: The Future Roles of Social Housing in England* CASEreport 34, ESRC Centre for Analysis of Social Exclusion, London School of Economics and Political Science,, London, viewed 8 June 2013, http://eprints.lse.ac.uk/5568/.

Hoekstra JSCM (2010) *Divergence in European Welfare and Housing Systems*, IOS Press, Amsterdam.

Hogwood BW & Gunn LA (1984) *Policy Analysis for the Real World*, Oxford University Press, Oxford.

Hollander J (1991) 'It All Depends', *Social Research*, 58, 31–49.

Holmans AE (1987) *Housing Policy in Britain: A History*, Croom Helm, London.

Housing Industry Association (2008–9) *Housing: The National Magazine for Building Professionals*, viewed 4 February 2010, http://hia.com.au/housing/.

Housing NSW (2007) *Planning for the Future: New Directions for Community Housing in New South Wales 2007/08–2012/13*, Housing NSW, Sydney, viewed 3 March 2011, http://www.housing.nsw.gov.au/NR/rdonlyres/EF5C1628-EB52-4DFE-9001-F4A31A5B6D61/0/PFTFCompleteDec2007.pdf.

——(2008) *Corporate Plan 2007/8–2009/10*, Housing NSW, Sydney, viewed 3 March 2011, http://www.housing.nsw.gov.au/NR/rdonlyres/A34AE89A-EA2D-46CD-84DD-27F9EC29117D/0/HousingNSWCorporatePlanexternal_web.pdf.

——(2011) *History of Public Housing in NSW*, viewed 25 October 2011, http://www.housing.nsw.gov.au/About+Us/History+of+Public+ Housing+in+NSW/.

Housing Theory and Society (2013) viewed 4 July 2013, http://www.tandfonline.com/action/aboutThisJournal?show=aimsScope&journalCode=shou20.

Howard E (1902) *Garden Cities of To-Morrow*, Swan Sonnenschein, London.

Howe R (ed.) (1988a) *New Houses for Old: Fifty Years of Public Housing in Victoria 1938–1988*, Ministry of Housing and Construction, Melbourne.

——(1988b) 'Reform and Social Responsibility: The Establishment of the Housing Commission', in R Howe (ed.) *New Houses for Old: Fifty Years of Public Housing in Victoria 1938–1988*, Ministry of Housing and Construction, Melbourne, pp. 20–44.

——(1988c) 'From Rehabilitation to Prevention: The War Years', in R Howe (ed.) *New Houses for Old: Fifty Years of Public Housing in Victoria 1938–1988*, Ministry of Housing and Construction, Melbourne, pp. 45–68.

Hulse K (2003) 'Housing Allowances and Private Renting in Liberal Welfare Regimes', *Housing, Theory and Society*, 20, 1, 28–42.

——(2008) 'Shaky Foundations: Moving Beyond 'Housing Tenure'', *Housing, Theory and Society*, 25, 3, 202–219>

Hulse K & Burke T (2005) *The Changing Role of Allocations Systems in Social Housing*, Final Report 75, AHURI, Melbourne, viewed 3 February 2010, http://www.ahuri.edu.au/publications/download/50141_fr.

Hulse K & Saugeres L (2008) *Home Life, Work and Housing Decisions: A Qualitative Analysis*, National Research Venture 1: Housing Assistance and Non-Shelter Outcomes: Research Paper 7, AHURI, Melbourne, viewed 21 June 2010, http://www.ahuri.edu.au/publications/download/nrv1_research_paper_7.

Hulse K & Stone W (2007) 'Social Cohesion, Social Capital and Social Exclusion: A Cross Cultural Comparison', *Policy Studies*, 28, 2, 109–128.

Hulse K, Herbert T & Down K (2004) *Kensington Estate Redevelopment Social Impact Study*, Swinburne University of Technology, Melbourne, viewed 26 September 2010, http://researchbank.swinburne.edu.au/vital/access/services/Download/swin: 242/DS3.

Hulse K, Neske C & Burke T (2006) *Improving Access to Social Housing: Ideas for Reform*, Positioning Paper 88, AHURI, Melbourne, viewed 3 February 2010, http://www.ahuri.edu.au/publications/download/50297_pp.

Hulse K, Phillips R & Burke T (2007) *Improving Access to Social Housing: Paradigms, Principles and Reforms*, Final Report 97, AHURI, Melbourne, viewed 3 February 2010, http://www.ahuri.edu.au/publications/download/50297_fr.

Husserl E (1970) *The Crisis of European Sciences and Transcendental Phenomenology: An Introduction to Phenomenological Philosophy*, Northwestern University Press, Evanston, Ill.

Industry Commission (1993a) *Public Housing: Volume 1: Report*, Australian Government Publishing Service, Canberra, viewed 4 February 2010, http://www.pc.gov.au/ic/inquiry/34public/34public.pdf.

——(1993b) *Public Housing: Volume 2: Appendices*, Australian Government Publishing Service, Canberra, viewed 4 February 2010, http://www.pc.gov.au/ic/inquiry/34public/34public.pdf.

Jacobs K (2001) 'Historical Perspectives and Methodologies: Their Relevance for Housing Studies?', *Housing, Theory and Society*, 18, 3–4, 127–135.

Jacobs K & Arthurson K (2003) *Developing Effective Housing Management Policies to Address Problems of Anti-Social Behaviour*, Final Report 50, AHURI, Melbourne, viewed 4 February 2010, http://www.ahuri.edu.au/publications/download/40163_fr.

Jacobs K & Manzi T (1996) 'Discourse and Policy Change: The Significance of Language for Housing Research', *Housing Studies*, 11, 4, 543–560.

——(2000) 'Evaluating the Social Constructionist Paradigm in Housing Research', *Housing, Theory and Society*, 17, 1, 35–42.

Jacobs K, Arthurson K, White R & Donoghue J (2003) *Developing Effective Housing Management Strategies to Address Problems of Anti-Social Behaviour*, Positioning Paper 54, AHURI, Melbourne, viewed 4 February 2010, http://www.ahuri.edu.au/publications/download/40163_pp.

Jacobs K, Atkinson R, Spinney A, Colic-Peisker V, Berry M & Dalton T (2010a) *What Future for Public Housing? A Critical Analysis*, Research Paper, AHURI, Melbourne, viewed 4 February 2010, http://www.ahuri.edu.au/publications/download/40561_rp.

Jacobs K, Atkinson R, Colic-Peisker V, Berry M & Dalton T (2010b) *What Future for Public Housing? A Critical Analysis*, Final Report 151, AHURI, Melbourne, viewed 9 June 2011, http://www.ahuri.edu.au/publications/download/40561_fr.

Jacobs K, Kemeny J & Manzi T (eds) (2004) *Social Constructionism in Housing Research*, Ashgate, Aldershot.

Jaspers K (1953) *The Origin and Goal of History*, Yale University Press, New Haven.

——(1954) *Way to Wisdom: An Introduction to Philosophy*, Yale University Press, New Haven.

Johnson L (1996) '"As Housewives We Are Worms": Women, Modernity and the Home Question', *Cultural Studies*, 10, 449–463.

Jones A & Seelig T (2004) *Understanding and Enhancing Research-Policy Linkages in Australian Housing: A Discussion Paper*, Positioning Paper 75, AHURI, Melbourne, viewed 18 July 2008, http://www.ahuri.edu.au/publications/download/20216_pp.

——(2005) *Enhancing Research-Policy Linkages in Australian Housing: An Options Paper*, Final Report 79, AHURI, Melbourne, viewed 18 July 2008, http://www.ahuri.edu.au/publications/download/20216_fr.

Jones MA (1972) *Housing and Poverty in Australia*, Melbourne University Press, Melbourne.

Judd B & Randolph B (2006) 'Qualitative Methods and the Evaluation of Community Renewal Programs in Australia: Towards a National Framework', *Urban Policy and Research*, 24, 1, 97–114.

Kemeny J (1981) *The Myth of Home Ownership: Private Versus Public Choices in Housing Tenure*, Routledge & Kegan Paul, London.

——(1983) *The Great Australian Nightmare: A Critique of the Home-Ownership Ideology*, Georgian House, Melbourne.

——(1987) 'Towards Theorised Housing Studies: A Counter-Critique of the Provision Thesis', *Housing Studies*, 2, 4, 249–260.

——(1988) 'Defining Housing Reality: Ideological Hegemony and Power in Housing Research', *Housing Studies*, 3, 4, 205–218.

——(1992) *Housing and Social Theory*, Routledge, London.

——(1995) *From Public Housing to the Social Market: Rental Policy Strategies in Comparative Perspective*, Routledge, London.

Kemeny J & Lowe S (1998) 'Schools of Comparative Housing Research: From Convergence to Divergence', Housing Studies, 13, 2, 161–176.

Kenley R, Chiazor M & Heywood C (2010) *Good Practices for Managing Australia's Public and Community Housing Assets*, Final Report 148, AHURI, Melbourne, viewed 9 February 2011, http://www.ahuri.edu.au/publications/download/50366_fr.

Kenley R, Chiazor M, Heywood C & McNelis S (2009) *Towards Best Practice for Public Housing Asset Management*, Positioning Paper 118, AHURI, Melbourne, viewed 4 February 2010, http://www.ahuri.edu.au/publications/download/50366_pp.

Kewley TH (1973) *Social Security in Australia, 1900–1972*, Sydney University Press, Sydney.

——(1980) *Australian Social Security Today: Major Developments from 1900 to 1978*, Sydney University Press, Sydney.

King P (1996) *The Limits of Housing Policy: A Philosophical Investigation*, Middlesex University Press, London.

——(1998) *Housing, Individuals and the State: The Morality of Government Intervention*, Routledge, London.

——(2003) *A Social Philosophy of Housing*, Ashgate, Aldershot.

——(2004) *Private Dwelling: Contemplating the Use of Housing*, Routledge, London.

——(2005) *The Common Place: The Ordinary Experience of Housing*, Ashgate, Aldershot.

——(2008) *In Dwelling: Implacability, Exclusion and Acceptance,* Ashgate, Farnham.

——(2009) 'Using Theory or Making Theory: Can There Be Theories of Housing?', *Housing, Theory and Society*, 26, 1, 41–52.

King RG (1993) 'Will the New Keynesian Macroeconomics Resurrect the IS–LM Model?', *Journal of Economic Perspectives*, 7, 1, 67–82.

Kleinman M & Whitehead C (1991) *Choosing a Rent Structure: A Handbook for Social Landlords*, Scottish Homes Technical Information Paper 2, Scottish Homes, Edinburgh.

Kliger B & McNelis S (2003a) *Frankston City Council: Social Housing Policy: Part I: Issues Paper*, Frankston City Council, Melbourne.

——(2003b) *Frankston City Council: Social Housing Policy: Part II: Action Plan*, unpublished.

Korpi W (1989) 'Power, Politics, and State Autonomy in the Development of Social Citizenship: Social Rights During Sickness in Eighteen Oecd Countries since 1930', *American Sociological Review*, 54, 3, 309–328.

KPMG (2005) *Evaluation of the Brisbane Housing Company: Executive Summary*, Queensland Department of Housing, Brisbane, viewed 7 December 2010, http://www.public-housing.qld.gov.au/design/pdf/bhc_evaluation_executive_ summary.pdf.

Kuhn TS (1970) *The Structure of Scientific Revolutions*, University of Chicago Press, Chicago.

Lakatos I (1970) 'Falsification and the Methodology of Scientific Research Programmes', in I Lakatos and A Musgrave (eds), *Criticism and the Growth of Knowledge: Proceedings of the International Colloquium in the Philosophy of Science*, Cambridge University Press, London.

——(1978) *The Methodology of Scientific Research Programmes*, Cambridge University, Cambridge.

Lamb ML (1982) *Solidarity with Victims: Toward a Theology of Social Transformation*, Crossroad, New York.

——(1985) 'The Dialectics of Theory and Praxis within Paradigm Analysis', *Lonergan Workshop*, 5, 71–114, viewed 17 February 2011, http://www.lonerganresource.com/pdf/journals/Lonergan_Workshop_Vol_5.pdf.

Lambert P & McShane P (2010) *Bernard Lonergan: His Life and Leading Ideas*, Axial, Vancouver.

Lawrence F (1978) 'The Horizon of Political Theology', in TA Dunne and J-M Laporte (eds) *Trinification of the World: A Festschrift in Honour of Frederick E Crowe in Celebration of His 60th Birthday*, Regis College Press, Toronto, pp. 46–70.

Lawrence RJ & Després C (2004) 'Futures of Transdisciplinarity', *Futures*, 36, 4, 397–405.

Lawrence RJ (2004) 'Housing and Health: From Interdisciplinary Principles to Transdisciplinary Research and Practice', *Futures*, 36, 4, 487–502.

Lawson JM (2001) 'Comparing the Causal Mechanisms Underlying Housing Networks Over Time and Space', *Journal of Housing and the Built Environment*, 16, 1, 29–52.

——(2002) 'Thin Rationality, Weak Social Constructionism and Critical Realism: The Way Forward in Housing Theory?', *Housing, Theory and Society*, 19, 3, 142–144.

——(2006) *Critical Realism and Housing Research*, Routledge, London.

——(2010) 'Path Dependency and Emergent Relations: Explaining the Different Role of Limited Profit Housing in the Dynamic Urban Regimes of Vienna and Zurich', *Housing, Theory and Society*, 27, 3, 204–220.

Lindberg DC (1992) *The Beginnings of Western Science: The European Scientific Tradition in Philosophical, Religious, and Institutional Context, 600 BC to AD 1450*, University of Chicago Press, Chicago.

Lloyd J & Johnson L (2004) 'Dream Stuff: The Postwar Home and the Australian Housewife 1940–60', *Environment and Planning D: Society and Space*, 22, 2, 251–272.

Lonergan BJF ([1942–1944] 1998) *For a New Political Economy* (Collected Works of Bernard Lonergan, Volume 21), University of Toronto Press, Toronto.

——([1957] 1992) *Insight: A Study of Human Understanding* (Collected Works of Bernard Lonergan, Volume 3), University of Toronto Press, Toronto.

——([1957] 2001a) 'Phenomenology: Nature, Significance, Limitations', in P McShane (ed.) *Phenomenology and Logic: The Boston College Lectures on Mathematical Logic and Existentialism* (Collected Works of Bernard Lonergan, Volume 18), University of Toronto Press, Toronto, pp. 266–279.

——([1957] 2001b) 'Subject and Horizon', in P McShane (ed.) *Phenomenology and Logic: The Boston College Lectures on Mathematical Logic and Existentialism* (Collected Works of Bernard Lonergan, Volume 18), University of Toronto Press, Toronto, pp. 280–297.

——([1957] 2001c) 'Horizon, History, Philosophy', in P McShane (ed.) *Phenomenology and Logic: The Boston College Lectures on Mathematical Logic and Existentialism* (Collected Works of Bernard Lonergan, Volume 18), University of Toronto Press, Toronto, pp. 298–317.

——([1959] 1993) *Topics in Education: The Cincinnati Lectures of 1959 on the Philosophy of Education* (Collected Works of Bernard Lonergan, Volume 10), University of Toronto Press, Toronto.

——([1959] 2013) 'Understanding and Method' in R Doran and HD Monsour (eds), *Early Works on Theological Method 2* (Collected Works of Bernard Lonergan, Volume 23) University of Toronto Press, Toronto.

——(1965) *Pages on 'Method' 697*, Bernard Lonergan Archive, unpublished, viewed 17 February 2010, http://www.bernardlonergan.com/pdf/69700DTE060.pdf.

——(1967) 'Cognitional Structure', in *Collection: Papers by Bernard JF Lonergan*, Darton, Longman & Todd, London, pp. 221–239.

——([1968] 1974) 'The Absence of God in Modern Culture', in W Ryan and B Tyrrell (eds) *A Second Collection: Papers by Bernard JF Lonergan, SJ*, Darton, Longman & Todd, London, pp. 101–116.

——([1968] 2010) '"Transcendental Philosophy and the Study of Religion", 3–12 July 1968, Boston College', in RM Doran and RC Croken (eds) *Early Works on Theological Method I* (Collected Works of Bernard Lonergan, Volume 22), University of Toronto Press, Toronto, pp. 421–633.

——(1969) 'Functional Specialties in Theology', *Gregorianum*, 50, 485–505.

——([1969]1974) 'The Future of Christianity', in W Ryan and B Tyrrell (eds) *A Second Collection: Papers by Bernard JF Lonergan, SJ*, Darton, Longman & Todd, London, pp. 149–163.

——([1972] 1990) *Method in Theology* (Collected Works of Bernard Lonergan, Volume 14), University of Toronto Press, Toronto.

——([1974] 1985) 'Second Lecture: Religious Knowledge', in FE Crowe (ed.) *A Third Collection: Papers by Bernard JF Lonergan*, Paulist Press, New York, pp. 129–145.

——(1974) 'Insight Revisited', in W Ryan and B Tyrrell (eds) *A Second Collection: Papers by Bernard JF Lonergan, SJ*, Darton, Longman & Todd, London, pp. 263–278.

——([1977] 1985) 'Natural Right and Historical Mindedness', in FE Crowe (ed.) *A Third Collection: Papers by Bernard JF Lonergan*, Paulist Press, New York, pp. 169–183.

——([1983] 1999) *Macroeconomic Dynamics: An Essay in Circulation Analysis* (Collected Works of Bernard Lonergan, Volume 15), University of Toronto Press, Toronto.

——(1997) *Verbum: Word and Idea in Aquinas* (Collected Works of Bernard Lonergan, Volume 2), University of Toronto Press, Toronto.

——(2000) *Grace and Freedom: Operative Grace in the Thought of St. Thomas Aquinas* (Collected Works of Bernard Lonergan, Volume 1), University of Toronto Press, Toronto.

——(2001) *Phenomenology and Logic: The Boston College Lectures on Mathematical Logic and Existentialism* (Collected Works of Bernard Lonergan, Volume 18), University of Toronto Press, Toronto.

Loorbach D & Rotmans J (2010) 'The Practice of Transition Management: Examples and Lessons from Four Distinct Cases', *Futures*, 42, 3, 237–246.

Lundqvist LJ (1988) 'Privatization: Towards a Concept for Comparative Policy Analysis', *Journal of Public Policy*, 8, 1, 1–19.

——(1992) *Dislodging the Welfare State?: Housing and Privatization in Four European Nations*, Delft University Press, Delft.

Maclennan D & Bannister J (1995) 'Housing Research: Making the Connections', *Urban Studies*, 32, 10, 1581–1586.

Maclennan D & More A (1997) 'The Future of Social Housing: Key Economic Questions', *Housing Studies*, 12, 4, 531–547.

Magarey S & Sheridan S (2002) 'Local, Global, Regional: Women's Studies in Australia', *Feminist Studies*, 28, 1, 129–152.

Mallett S (2004) 'Understanding Home: A Critical Review of the Literature', *Sociological Review*, 52, 1, 62–89.

Malpass P (1990) *Reshaping Housing Policy: Subsidies, Rents and Residualisation*, Routledge, London.

——(2000) 'The Discontinuous History of Housing Associations in England', *Housing Studies*, 15, 2, 195–212.

——(2005) *Housing and the Welfare State: The Development of Housing Policy in Britain*, Houndmills, Basingstoke.

Marcuse H (1968) *Negations: Essays in Critical Theory*, Allen Lane, London.

Marsden S (1986) *Business, Charity and Sentiment: The South Australian Housing Trust, 1936–1986*, Wakefield, Adelaide.

Marston G (2000) 'Metaphor, Morality and Myth: A Critical Discourse Analysis of Public Housing Policy in Queensland', *Critical Social Policy*, 20, 3, 349–373.

——(2002) 'Critical Discourse Analysis and Policy-Orientated Housing Research', *Housing, Theory & Society*, 19, 2, 82–91.

——(2004) 'Managerialism and Public Housing Reform', *Housing Studies*, 19, 1, 5–20.

——(2008) 'Technocrats or Intellectuals? Reflections on the Role of Housing Researchers as Social Scientists', *Housing, Theory and Society*, 25, 3, 177–190.

Marx K ([1845] 2002) *Theses on Feuerbach*, viewed 17 February 2010, http://www.marxists.org/archive/marx/works/1845/theses/index.htm.

Mathews WA (1987) 'Explanation in the Social Sciences', in TP Fallon and PB Riley (eds) *Religion and Culture: Essays in Honor of Bernard Lonergan, SJ*, State University of New York Press, Albany, pp. 245–260.

——(1998) 'A Biographical Perspective on Conversion and the Functional Specialties in Lonergan', *Method: A Journal of Lonergan Studies*, 16, 2, 133–160, viewed 10 October 2011, http://www.lonerganresource.com/pdf/journals/Method_Vol_16_No_2.pdf.

——(2005) *Lonergan's Quest: A Study of Desire in the Authoring of Insight*, University of Toronto Press, Toronto.

Mazur C & Wilson E (2011) *2010 Census Briefs: Housing Characteristics*, viewed 12 April 2013, http://www.census.gov/prod/cen2010/briefs/c2010br-07.pdf.

McNelis S (1992) *Social Housing: A Future Direction for Housing*, Social Housing Paper 1, Victorian Council of Social Service, Melbourne.

——(1993) *Social Housing: Building the Future*, Social Housing Paper 2, Victorian Council of Social Service, Melbourne.

——(1996) *Performance Monitoring and Housing Assistance in Victoria*, Victorian Council of Social Service, Melbourne.

——(2000) *Ideology and Public Housing Rental Systems: A Case Study of Public Housing in Victoria*, Master of Arts thesis, Swinburne University of Technology, Melbourne.

——(2004) *Independent Living Units: The Forgotten Social Housing Sector*, Final Report, AHURI, Melbourne, viewed 13 December 2011, http://www.ahuri.edu.au/publications/download/50138_fr.

——(2005) 'Rental Policy: Financial Viability or Affordability in Australian Public Housing', *Building for Diversity: National Housing Conference 2005*, AHURI and Department of Housing and Works, Perth, pp. 83–108.

——(2006) *Rental Systems in Australia and Overseas*, Final Report, AHURI, Melbourne, viewed 4 February 2010, http://www.ahuri.edu.au/publications/download/50226_fr.

——(2007) *Older Persons in Public Housing: Present and Future Profile,* Research Report, AHURI, Melbourne, viewed 4 February 2010, http://www.ahuri.edu.au/publications/download/50318_rp.

——(2009a) 'Rent-Setting in Social Housing: An Interpretation in the Context of Global Functional Collaboration', *Halifax Lonergan Conference 2009: Global Functional Collaboration*, Saint Mary's University, Halifax, NS, 6–10 July.

——(2009b) 'Comparative Housing Research and Policy: Social Housing Rent-Setting in Western Countries', *International Sociological Association International Housing Conference 2009: Housing Assets, Housing People*, University of Glasgow, Glasgow, 1–4 September, viewed 15 July 2010, http://www.gla.ac.uk/media/media_129713_en.pdf.

——(2010) 'A Prelude to (Lonergan's) Economics', *The Lonergan Review*, 11, 1, 107–120.

McNelis S & Burke T (2004) *Rental Systems in Australia and Overseas*, Positioning Paper, AHURI, Melbourne, viewed 4 February 2010, http://www.ahuri.edu.au/publications/download/50226_pp.

McNelis S & Herbert T (2003) *Independent Living Units: Clarifying Their Current and Future Role as an Affordable Housing Option for Older Persons with Low Assets and Low Incomes*, Positioning Paper 59, AHURI, Melbourne, viewed 20 October 2011, http://www.ahuri.edu.au/publications/download/50138_pp.

McNelis S & Nichols R (1997) *Towards a Holistic Response to People with Housing and Support Needs: A Report to the Office of Housing on Integrating Housing Assistance and Support Services for People with Disabilities*, Victorian Council of Social Service, Melbourne.

McNelis S & Reynolds A (2001) *Creating Better Futures for Residents of High-Rise Public Housing in Melbourne,* Ecumenical Housing, Melbourne, viewed 19 January 2011, http://researchbank.swinburne.edu.au/vital/access/services/Download/swin: 2970/SOURCE1.

McNelis S, Burke T & Neske C (2006) *Affordable Housing Discussion Paper: City of Yarra*, City of Yarra, Melbourne.

McNelis S, Esposto A & Neske C (2005a) *Affordable Housing for Young People Employed in the City of Melbourne: Volume I Summary Report*, City of Melbourne, Melbourne.

——(2005b) *Affordable Housing for Young People Employed in the City of Melbourne: Volume II Technical Report*, City of Melbourne, Melbourne.

McNelis S, Hayward D & Bisset H (2001) *A Private Retail Investment Vehicle for the Community Housing Sector*, Positioning Paper, AHURI, Melbourne, viewed 13 December 2011, http://www.ahuri.edu.au/publications/download/50022_pp.

——(2002a) *A Private Retail Investment Vehicle for the Community Housing Sector*, Work-in-Progress Report, AHURI, Melbourne.

——(2002b) *A Private Retail Investment Vehicle for the Community Housing Sector*, Final Report, AHURI, Melbourne, viewed 13 December 2011, http://www.ahuri.edu.au/publications/download/50022_fr.

McNelis S, Kliger B, Burke T, Ralston L & Neske C (2005a) *Bass Coast Shire Local Housing Picture,* Shire of Bass Coast, Wonthaggi, viewed 13 December 2011, http://researchbank.swinburne.edu.au/vital/access/services/Download/swin: 1709/DS2.

——(2005b) *Bass Coast Shire Affordable Housing Strategy*, Shire of Bass Coast, Wonthaggi.

McNelis S, Kliger B, Nankervis M, Burke T, Ralston L & Neske C (2005c) *Shire of Yarra Ranges Housing Strategy Issues Paper*, Shire of Yarra Ranges, viewed 3 March 2011, http://researchbank.swinburne.edu.au/vital/access/services/Download/ swin: 1729/DS2.

McNelis S, Neske C, Jones A & Phillips R (2008) *Older Persons in Public Housing: The Policy and Management Issues*, Final Report 121, AHURI, Melbourne, viewed 22 June 2010, http://www.ahuri.edu.au/publications/download/50318_fr.

McNelis S, O'Brien A, Reynolds A & McVicar G (2002) *Young Adults with Disabilities: Planning for Their Future Housing and Support*, Ecumenical Housing, Melbourne.

McShane P (1974) *Wealth of Self and Wealth of Nations: Self-Axis of the Great Ascent*, viewed 14 July 2010, http://www.philipmcshane.ca/wealth.pdf.

——(1976) *The Shaping of the Foundations: Being at Home in the Transcendental Method*, University Press of America, Washington, viewed 12 February 2010, http://www.philipmcshane.ca/foundations.pdf.

——(1980) *Lonergan's Challenge to the University and the Economy,* University Press of America, Washington, viewed 12 February 2010, http://www.philipmcshane.ca/lonerganschallenge.pdf.

——(1984) 'Middle Kingdom: Middle Man', in P McShane (ed.) *Searching for Cultural Foundations*, University Press of America, New York, pp. 1–43.

——(1989) *Process: Introducing Themselves to Young (Christian) Minders,* viewed 4 March 2010, http://www.philipmcshane.ca/process.pdf.

——(1998) *Economics for Everyone: Das Jus Kapital*, Axial, Halifax.

——(2002) *Pastkeynes Pastmodern Economics: A Fresh Pragmatism*, Axial, Halifax.

——(2003) *Cantower 15: The Elements of Meaning*, viewed 8 July 2011, http://www.philipmcshane.ca/cantower15.pdf.

——(2004a) *Sofdaware 1–8*, viewed 8 September 2011, http://www.philipmcshane.ca/sofdaware.html.

——(2004b) *Quodlibet 1–21*, viewed 8 September 2011, http://www.philipmcshane.ca/quodlibet.html.

——(2004c) *Sofdaware 6: Rambles in Method 250*, viewed 3 November 2010, http://www.philipmcshane.ca/sofda-06.pdf.

——(2004d) *Quodlibet 10: A Simple Dialectic Positioning on Functional Specialization*, viewed 16 February 2010, http://www.philipmcshane.ca/quod-10.pdf.

——(2005a) 'The Origins and Goals of Functional Specialization', Twentieth Annual Fallon Memorial Lonergan Symposium, 31 March – 2 April, Loyola Marymount University, Los Angeles http://lonergan.concordia.ca/reprints/mcshane/McShane_Functional_Specialization_2004.pdf.

——(2007a) *Method in Theology: Revisions and Implementations*, viewed 12 February 2010, http://www.philipmcshane.ca/method.html.

——(2007b) *Lonergan's Standard Model of Effective Global Enquiry*, viewed 16 February 2010, http://www.philipmcshane.ca/lonergansmodel.html.

——(2007c) *Prehumous 1: Teaching High School Economics. A Common-Quest Manifesto*, viewed 12 February 2010, http://www.philipmcshane.ca/prehumous-01.pdf.

——(2008) *Surf 5: Cosmopolis and Functional Differentiations*, viewed 16 February 2010, http://www.philipmcshane.ca/SURF-05.pdf.

——(2009) *Fusion 5: What Collaboration Might Be Achieved in 2010–2015?*, viewed 16 February 2010, http://www.philipmcshane.ca/fusion-05.pdf.

Mead GH (1938) *The Philosophy of the Act*, University of Chicago Press, Chicago.

Melchin KR (1991) 'Moral Knowledge and the Structure of Co-operative Living', *Theological Studies*, 52, 3, 495–523.

——(1994) 'Economies, Ethics, and the Structure of Social Living', *Humanomics*, 10, 3, 21–57.

——(1998) *Living with Other People: An Introduction to Christian Ethics Based on Bernard Lonergan*, Novalis, Ottawa.

——(1999) *History, Ethics, and Emergent Probability: Ethics, Society, and History in the Work of Bernard Lonergan*, The Lonergan Web Site, Toronto, viewed 6 June 2013, http://www.loneranresource.com/pdf/books/6/Melchin,_Kenneth_-_History,_Ethics,_and_Emergent_Probability.pdf.

——(2003) 'Exploring the Idea of Private Property: A Small Step Along the Road from Common Sense to Theory', *Journal of Macrodynamic Analysis*, 3, 287–301, viewed 6 June 2013, http://journals.library.mun.ca/ojs/index.php/jmda/article/view/131/84.

Merrett S (1979) *State housing in Britain*, Routledge & Kegan Paul, London

Meynell HA (1975) 'Lonergan's Theory of Knowledge and the Social Sciences', *New Blackfriars*, 56, 388–398.

——(1998) *Redirecting Philosophy: Reflections on the Nature of Knowledge from Plato to Lonergan*, University of Toronto Press, Toronto.

Mill JS ([1882] 2009) *A System of Logic, Ratiocinative and Inductive, Being a Connected View of the Principles of Evidence, and the Methods of Scientific Investigation,* Project Gutenberg, viewed 9 March 2011, http://www.gutenberg.org/ebooks/27942.

Milligan V (2003) 'How Different? Comparing Housing Policies and Housing Affordability Consequences for Low Income Households in Australia and the Netherlands', *Nederlandse Geografische Studies*, 318, 19–246.

Milligan V, Gurran N, Lawson J, Phibbs P & Phillips R (2009) *Innovation in Affordable Housing in Australia: Bringing Policy and Practice for Not-for-Profit Housing Organisations Together*, Final Report 134, AHURI, Melbourne, viewed 7 October 2011, http://www.ahuri.edu.au/publications/download/60504_fr.

Milligan V, Phibbs P, Gurran N & Fagan K (2007) *Approaches to Evaluation of Affordable Housing Initiatives in Australia,* National Research Venture 3: Housing Affordability for Lower Income Australians, Research Paper 7, AHURI, Melbourne, viewed 26 September 2010, http://www.ahuri.edu.au/publications/download/nrv3_research_paper_7.

Mintzberg H (ed.) (2007) *Tracking Strategies: Toward a General Theory*, Oxford University Press, Oxford.

Moore S (1989) *Jesus the Liberator of Desire*, Crossroad, New York.

Morelli MD (1995) 'The Polymorphism of Human Consciousness and the Prospects for a Lonerganian History of Philosophy', *International Philosophical Quarterly*, 35, 4, 379–402.

Mullins D (2006) *Housing Policy in the UK*, Palgrave Macmillan, Basingstoke.

Næss A (1988) 'Deep Ecology and Ultimate Premises', *Ecologist*, 18, 4/5, 128–131.

National Housing Policy Review (1989) *Background Paper*, Department of Community Services and Health, Canberra.

National Housing Strategy (1992) *National Housing Strategy: Agenda for Action*, Australian Government Publishing Service, Canberra.

National Shelter (2001) *The Way Forward: Affordable Housing for All*, National Shelter, Canberra.

——(2004) *Rebuilding the Australian Dream: A National Approach: National Shelter Policy Platform 2004*, National Shelter, Canberra.

——(2009) *Housing Australia Affordably: National Shelter Policy Platform*, viewed 4 February 2010, http://www.shelter.org.au/NS%20Policy%20Platform%20 2009%20 long%20final.pdf.

Neuman WL (2006) *Social Research Methods: Qualitative and Quantitative Approaches*, Pearson, Boston.

O'Leary DM (1998) *Lonergan's Practical View of History*, MA thesis, Regis College, University of Toronto, Toronto.

O'Neill P (2008) 'The Role of Theory in Housing Research: Partial Reflections of the Work of Jim Kemeny', *Housing, Theory and Society*, 25, 3, 164–176.

Oliver P (2003) *Dwellings: The Vernacular House World Wide*, Phaidon, London.

Ortega y Gasset J ([1946] 1998) *Mission of the University*, Routledge, Florence.

Oxley M (1989) 'Housing Policy: Comparing International Comparisons', *Housing Studies*, 4, 2, 128–132.

——(1991) 'The Aims and Methods of Comparative Housing Research', *Scandinavian Housing and Planning Research*, 8, 2, 67–77.

——(2001) 'Meaning, Science, Context and Confusion in Comparative Housing Research', *Journal of Housing and the Built Environment*, 16, 1, 89–106.

Oyler D (2010) *SGEME 8: Generalizing Functional Specialization*, viewed 8 July 2011, http://www.sgeme.org/Articles/sgeme-008-oyler-wcmi-2010.pdf.

Paris C, Beer AP & Sanders W (1993) *Housing Australia*, Macmillan Education, South Melbourne.

Parsons DW (1995) *Public Policy: An Introduction to the Theory and Practice of Policy Analysis*, Edward Elgar, Aldershot.

Parsons T ([1951] 1991) *The Social System*, Taylor & Francis, Hoboken.

Patton MQ (2002) *Qualitative Research and Evaluation Methods*, Sage, Thousand Oaks.

Pawson R (2002) 'Evidence-Based Policy: The Promise of 'Realist Synthesis'', *Evaluation*, 8, 3, 340–358.

Peel M (1995) *Good Times, Hard Times: The Past and the Future in Elizabeth*, Melbourne University Press, Melbourne.

Phibbs P & Thompson S (2011) *The Health Impacts of Housing: Toward a Policy-Relevant Research Agenda*, Final Report 173, AHURI, Melbourne, viewed 24 October 2011, http://www.ahuri.edu.au/publications/download/70619_fr.

Planning Institute Australia (2007) *National Position Statement: Affordable Housing*, Planning Institute Australia, viewed 4 February 2010, http://www.planning.org.au/ index.php?option=com_docman&task=docclick&Itemid=0&bid=2312&limitstart=0&l imit=10.

Plato ([360 BC] 1892) *The Republic,* trans. B. Jowett, Floating Press, Auckland, viewed 18 February 2010, http://www.gutenberg.org/etext/1497.

Popper K (2002) *The Logic of Scientific Discovery,* Routledge, London.

Priorities Review Staff (1975) *Report on Housing,* Australian Government Publishing Service, Canberra.

Productivity Commission (2010a) *Strengthening Evidence Based Policy in the Australian Federation: Roundtable Proceedings, Volume 1: Proceedings,* Productivity Commission, Canberra, viewed 22 February 2011, http://www.pc.gov.au/research/confproc/strengthening-evidence.

——(2010b) *Strengthening Evidence Based Policy in the Australian Federation: Roundtable Proceedings, Volume 2: Background Paper,* Productivity Commission, Canberra, viewed 22 February 2011, http://www.pc.gov.au/research/confproc/strengthening-evidence.

Pugh C (1976) *Intergovernmental Relations and the Development of Australian Housing Policies,* Research Monograph 15, Centre for Research on Federal Financial Relations, Australian National University, Canberra.

Quinn TJ (2005) 'Reflections on Progress in Mathematics (Method in Mathematics)', West Coast Methods Institute Conference, Los Angeles, unpublished.

——(2012) 'Invitation to Functional Collaboration: Dynamics of Progress in the Sciences, Technologies, and Arts' *Journal of Macrodynamic Analysis,* viewed 14 February 2012, http://journals.library.mun.ca/ojs/index.php/jmda/article/download/362/234.

Ravetz A (2001) *Council housing and Culture: The History of a Social Experiment,* Routledge, London.

Raymaker JA (1977) *Theory-Praxis of Social Ethics: The Complementarity Between Bernard Lonergan's and Gibson Winter's Theological Foundations,* PhD thesis, Faculty of the Graduate School, Marquette University, Milwaukee.

Read P (ed.) (2000) *Settlement: A History of Australian Indigenous Housing,* Aboriginal Studies Press, Canberra.

Reynolds A, Bigby C & McNelis S (1999) *Accommodation Options for People Ageing with a Disability,* Ecumenical Housing, Melbourne.

Ricoeur P (1970) *Freud and Philosophy: An Essay on Interpretation,* Yale University Press, New Haven.

Ritzer G (ed.) (2003) *The Blackwell Companion to Major Classical Social Theorists,* Blackwell, Malden.

——(2005) *Encyclopedia of Social Theory,* Sage, Thousand Oaks.

Robinson J (1962) *Economic Philosophy,* Penguin, London.

Rogers P, Stevens K, Briskman L & Berry M (2005) *Framework for Evaluating Building a Better Future: Indigenous Housing to 2010,* Final Report 82, vols 1, 2 and 3 (Appendix 6), AHURI, Melbourne, viewed 7 October 2011, http://www.ahuri.edu.au/publications/projects/p30235.

Romer D (2000) 'Keynesian Macroeconomics Without the LM Curve', *Journal of Economic Perspectives,* 14, 2, 149–169.

Ronald R (2008) *The Ideology of Home Ownership: Homeowner Societies and the Role of Housing,* Palgrave Macmillan, Basingstoke.

Rorty R (1989) *Contingency, Irony, and Solidarity,* Cambridge University, Cambridge.

Roy Morgan Research (2007) *2007 National Social Housing Survey: Public Housing National Report,* Australian Institute of Health and Welfare, Canberra, viewed 8 June 2010, http://www.aihw.gov.au/housing/assistance/nshs/nshs_2007/2007_nshs_pha_including_appendix_1.pdf.

Ruming K (2006) *MOSAIC Urban Renewal Evaluation Project: Urban Renewal Policy, Program and Evaluation Review*, City Futures Research Centre, University of NSW, Sydney, viewed 18 October 2013, https://www.be.unsw.edu.au/sites/default/files/upload/research/centres/cf/publications/researchpapers/researchpaper4.pdf.

Sauer JB (1995) 'Economics and Ethics: Foundations for a Transdisciplinary Dialogue', *Humanomics*, 11, 1–2, 5–91.

Saugeres L (1999) 'The Social Construction of Housing Management Discourse: Objectivity, Rationality and Everyday Practice', *Housing, Theory and Society*, 16, 3, 93–105.

Sayer A (2010) *Method in Social Science: A Realist Approach*, 2nd edn, Routledge, London.

Schram S & Caterino B (eds) (2006) *Making Political Science Matter: Debating Knowledge, Research, and Method*, New York University Press, New York.

Schutz A (1972) *The Phenomenology of the Social World*, Heinemann Educational, London.

——(1982) *The Problem of Social Reality: Collected Papers Volume 1*, Martinus Nijhoff, The Hague.

Scottish Housing Regulator (2013) *Statistical Information*, viewed 28 March 2013, http://www.scottishhousingregulator.gov.uk/find-and-compare-landlords/statistical-information.

SCRGSP (2012) *Report on Government Services 2012 – Errata Chapter 16 with attachment tables, and Sector summary*, Steering Committee for the Review of Government Service Provision, Productivity Commission, Canberra, viewed 10 November 2012, http://www.pc.gov.au/__data/assets/pdf_file/0010/115768/government-services-2012-chapter16-sectorg-errata.pdf.

Scriven M (1991) *Evaluation Thesaurus*, Sage, Newbury Park.

Shelter England (2013) *Policy Library*, viewed 8 June 2013, http://england.shelter.org.uk/professional_resources/policy_and_research/policy_library.

Shute M (1994) 'Emergent Probability and the Ecofeminist Critique of Hierarchy', in CSW Crysdale (ed.) *Lonergan and Feminism*, University of Toronto Press, Toronto, pp. 146–174.

——(2010a) *Lonergan's Discovery of the Science of Economics*, University of Toronto Press, Toronto.

——(ed.) (2010b) *Lonergan's Early Economic Research: Texts and Commentary*, University of Toronto Press, Toronto.

——(2010c) 'Real Economic Variables', *Divyadaan: Journal of Philosophy and Education*, 21, 2.

——(2010d) 'Two Fundamental Notions of Economic Science', *The Lonergan Review*, 2, 1.

——(2013a) 'Functional Collaboration as the Implementation of 'Lonergan's Method': Part 1: For What Problem is Functional Collaboration the Solution?', *Divyadaan: Journal of Philosophy & Education*, 24, 1, 1–34.

——(2013b) 'Functional Collaboration as the Implementation of Lonergan's Method: Part 2: How Might We Implement Functional Collaboration?', *Divyadaan: Journal of Philosophy & Education*, 24, 2, 159–190.

Shute M & Zanardi W (2006) *Improving Moral Decision-Making*, McGraw Hill, Boston.

Sibley J, Hes D & Martin F (2003) 'Triple Helix Research: An Inter-Disciplinary Approach to Research into Sustainability in Outer-Suburban Housing Estates', Methodologies in Housing Research Conference, Stockholm.

Silverman, R M & Patterson, K L (2012) 'The Four Horsemen of the Fair Housing Apocalypse: A Critique of Fair Housing Policy in the USA', *Critical Sociology*, 38, 1, 123–140.

Simmel G ([1918] 2010) *The View of Life: Four Metaphysical Essays with Journal Aphorisms*, University of Chicago Press, Chicago.

Smith A ([1776] 2009) *An Inquiry into the Nature and Causes of the Wealth of Nations*, Project Gutenberg, viewed 9 February 2011, http://www.gutenberg.org/etext/3300.

Snell B (1953) *The Discovery of the Mind: The Greek Origins of European Thought*, Blackwell, Oxford.

Somerville P & Bengtsson B (2002) 'Constructionism, Realism and Housing Theory', *Housing, Theory and Society*, 19, 3, 121–136.

South Australia Building South Australia (2005) *Housing Plan for South Australia*, Adelaide, viewed 3 March 2011, http://blogs.dfc.sa.gov.au/m/dfcweb_hsa/ 501/ download.aspx.

Spiller Gibbins Swan (2000) *Public Housing Estate Renewal in Australia*, Australian Housing Research Fund Project 212, Spiller Gibbins Swan, Melbourne.

St Amour P (2009) 'Situating Lonergan's Economics in a Context of Collaboration', Lonergan Workshop, Boston, 21–26 June, unpublished.

Tegart G (2003) 'Technology Foresight: Philosophy & Principles', *Innovation: Management, Policy & Practice*, 5, 2–3, 279–285.

Tenants Union of Victoria (2010) viewed 4 Feburary 2010, http://www.tuv.org.au/advice/ resources.aspx.

Thompson B (2002) 'Multiracial Feminism: Recasting the Chronology of Second Wave Feminism', *Feminist Studies*, 28, 2, 337–360.

Tibbits G (1988) '"The Enemy within Our Gates": Slum Clearance and High-Rise Flats', in R Howe (ed.) *New Houses for Old: Fifty Years of Public Housing in Victoria 1938–1988*, Ministry of Housing and Construction, Melbourne, pp. 123–162.

Travers M, Gilmour T, Jacobs K, Milligan V & Phillips R (2011) *Stakeholder Views of the Regulation of Affordable Housing Providers in Australia* Final Report 161, AHURI, Melbourne, viewed 8 June 2013, http://www.ahuri.edu.au/publications/download/ 40559_fr.

Troy PN (1992) 'The Evolution of Government Housing Policy: The Case of New South Wales 1901–41', *Housing Studies*, 7, 3, 216–233.

——(2000) *A History of European Housing in Australia*, Cambridge University Press, Cambridge.

——(2012) *Accommodating Australians : Commonwealth Government Involvement in Housing*, Federation Press, Annandale.

Tsenkova S (1996) 'Bulgarian Housing Reform and Forms of Housing Provision', *Urban Studies*, 33, 7, 1205–1219.

Turner JFC (1976) *Housing by People: Towards Autonomy in Building Environments*, Marion Boyars, London.

UK Department of Communities and Local Government (2010) *Evaluation of the Mixed Communities Initiative Demonstration Projects: Final Report*, viewed 3 May 2013, https: //www.gov.uk/government/uploads/system/uploads/attachment_data/file/6360/ 1775216.pdf.

——(2012) *Local Authority Housing Statistics: 2011–12: Local Authority-Owned Stock and Stock Management*, viewed 13 April 2013, https: //www.gov.uk/government/ uploads/system/uploads/attachment_data/file/39457/Local_authority_housing_ statistics_2011_12_v4.pdf.

UK Department of the Environment, Transport and the Regions (2000a) *Quality and Choice: A Decent Home for All: The Housing Green Paper*, Department of the Environment, Transport and the Regions, London, viewed 14 July 2013, http://webarchive.nationalarchives.gov.uk/20120919132719/http://www.communities.gov.uk/documents/housing/pdf/138019.pdf.

——(2000b) *Quality and Choice: A Decent Home for All: The Way Forward for Housing*, Department of the Environment, Transport and the Regions, London, viewed 14 July 2013, http://webarchive.nationalarchives.gov.uk/20120919132719/http://www.communities.gov.uk/documents/housing/pdf/138028.pdf.

UK House of Commons. Communities and Local Government Committee (2012) *Financing of New Housing Supply* HC 1652, House of Commons, London, viewed 8 June 2013, http://www.publications.parliament.uk/pa/cm201012/cmselect/cmcomloc/1652/1652.pdf.

UK Housing Corporation (2008) *721 Housing Quality Indicators (HQI) Form, Homes and Communities Agency*, viewed 24 July 2013, http://webarchive.nationalarchives.gov.uk/20100403094322/http://www.housingcorp.gov.uk/upload/pdf/721_HQI_Form_4_Apr_08_update_20080820153028.pdf.

UK Office for National Statistics (2013) *2011 Census*, viewed 12 April 2013, http://www.ons.gov.uk/ons/guide-method/census/2011/census-data/index.html.

van Overmeeren A & Gruis V (2011) 'Asset Management of Social Landlords Based on Value Creation at Neighbourhood Level', *Property Management*, 29, 2, 181–194.

VCOSS (1989) *VCOSS Response to the Proposed 1989 Commonwealth-State Housing Agreement* (authored by Sean McNelis), Victorian Council of Social Service, Melbourne.

——(1990) *VCOSS Housing Budget Submission to the Victorian Government, 1990–91* (authored by Sean McNelis), Victorian Council of Social Service, Melbourne.

——(1991a) *VCOSS Housing Budget Submission to the Victorian Government, 1991–92* (authored by Sean McNelis), Victorian Council of Social Service, Melbourne.

——(1991b) *On Foundations, Frameworks and Dreams: A Response to Issues Paper 2 and Issues Paper 3 of the National Housing Strategy* (authored by Sean McNelis), Victorian Council of Social Service, Melbourne.

——(1992) *VCOSS Housing Budget Submission to the Victorian Government, 1992–93* (authored by Sean McNelis), Victorian Council of Social Service, Melbourne.

——(1993a) *VCOSS Submissions to the Industry Commission Inquiry into Public Housing: Preliminary Submission and Response to Draft Paper* (authored by Sean McNelis), Victorian Council of Social Service, Melbourne.

——(1993b) *VCOSS Response to the Draft Victorian Community Housing Plan* (authored by Sean McNelis), Victorian Council of Social Service, Melbourne.

——(2008) *Universal Housing Universal Benefits: A VCOSS Discussion Paper on Universal Housing Regulation in Victoria,* Victorian Council of Social Service, Melbourne, viewed 4 February 2010, http://www.vcoss.org.au/documents/VCOSS%20docs/Housing/VUHA/Universal%20Housing%20Universal%20Benefits-email.pdf.

Vico G ([1744] 1948) *The New Science of Giambattista Vico*, translated from the third edition by Thomas Goddard Bergin and Max Harold Fisch, Cornell University Press, New York, viewed 6 June 2013, http://archive.org/download/newscienceofgiam030174mbp/newscienceofgiam030174mbp.pdf.

Victoria Housing and Community Building (2004) *Partnerships for Better Housing Assistance: Housing and Community Building's Strategic Framework 2004–09*, Department of Human Services, Melbourne.

Victoria Ministry of Housing (1984) *Proposals for the Implementation of the CSHA Cost Rent Formula in Victoria: Discussion Paper*, Ministry of Housing, Melbourne.

Victorian Office of Housing (2006–2010) *Debt Management Manual*, Office of Housing, Department of Human Services, Melbourne, viewed 28 October 2011, http://www.dhs. vic.gov.au/about-the-department/documents-and-resources/policies,-guidelines-and-legislation/debt-management-manual.

——(2008–2010b) *Housing Standards Policy Manual*, Office of Housing, Department of Human Services, Melbourne, viewed 28 October 2011, http://www.dhs.vic.gov.au/ about-the-department/documents-and-resources/policies,-guidelines-and-legislation/ housing-standards-policy-manual.

——(2009) *Rental Rebate Policy & Procedures Manual, Version 4.0, May 2009*, Office of Housing, Department of Human Services, Melbourne, viewed 28 October 2011, http:// www.dhs.vic.gov.au/about-the-department/documents-and-resources/policies,-guidelines-and-legislation/rental-rebate-manual.

——(2010) *Allocations Policy and Procedure Manual, Version 6.0, September 2010*, Office of Housing, Department of Human Services, Melbourne, viewed 28 October 2011, http://www.dhs.vic.gov.au/about-the-department/documents-and-resources/policies, -guidelines-and-legislation/allocations-manual.

Voros J (2003) 'A Generic Foresight Process Framework', *Foresight*, 5, 3, 10–21.

Wacquant L (2004) 'Critical Thought as Solvent of Doxa', *Constellations: An International Journal of Critical & Democratic Theory*, 11, 1, 97–101.

Walker R, Ballard J & Taylor C (2002) *Investigating Appropriate Evaluation Methods and Indicators for Indigenous Housing Programs*, Positioning Paper 24, AHURI, Melbourne, viewed 7 December 2010, http://www.ahuri.edu.au/publications/download/80037_pp.

——(2003) *Developing Paradigms and Discourses to Establish More Appropriate Evaluation Frameworks and Indicators for Housing Programs*, Final Report 29, AHURI, Melbourne, viewed 7 December 2010, http://www.ahuri.edu.au/publications/ download/80037_fr.

Walker RM (2000) 'The Changing Management of Social Housing: The Impact of Externalisation and Managerialisation', *Housing Studies*, 15, 2, 281–299.

Watson S (1986) 'Women and Housing or Feminist Housing Analysis', *Housing Studies*, 1, 1, 1–10.

——(1988a) *Accommodating Inequality: Gender and Housing*, Allen & Unwin, Sydney.

——(1988b) *The State or the Market? The Impact of Housing Policy on Women in New Zealand*, Royal Commission on Social Policy, Wellington.

Watson S & Austerberry H (1986) *Housing and Homelessness: A Feminist Perspective*, Routledge and Kegan Paul, London.

Wellek R & Warren A (1963) *Theory of Literature*, Penguin, Harmondsworth.

Winch P (2003) *The Idea of a Social Science and Its Relation to Philosophy*, Taylor & Francis, e-Library.

Winter G (1966) *Elements for a Social Ethic: Scientific Perspectives on Social Process*, Macmillan, New York.

Winter I & Seelig T (2001) 'Housing Research, Policy Relevance and a Housing Imagination in Australia', 2001 Housing Studies Association Conference, Cardiff, 4–5 September.

Wright B (2000) *Cornerstone of the Capital: A History of Public Housing in Canberra*, ACT Housing, Canberra.

Wright Mills C (1970) *The Sociological Imagination*, Penguin, Harmondsworth.

Yates J (*c*.1994) *Strategic Directions for Reforms of Public Housing Rent Policy: A Response to the Industry Commission's Report on Public Housing*, unpublished.

Yates J, Milligan V, Berry M, Burke T, Gabriel M, Phibbs P, Pinnegar S & Randolph B (2007) *Housing Affordability: A 21st Century Problem*, Final Report 105, AHURI, Melbourne, viewed 12 June 2010, http://www.ahuri.edu.au/publications/download/nrv3_final_report.

Young K, Ashby D, Boaz A & Grayson L (2002) 'Social Science and the Evidence-Based Policy Movement', *Social Policy and Society*, 1, 3, 215–224.

Index

Page references containing 'n' refer to notes.

Printed and bound by CPI Group (UK) Ltd, Croydon, CR0 4YY

23/10/2024

01778221-0008